TABLE OF CONTENTS

ACRONYMS

CNG	California National Guard
COP	Common Operational Picture
DCE	Defense Coordinating Element
DCO	Defense Coordinating Officer
DOD	Department of Defense
DSCA	Defense Support to Civil Authorities
EMAC	Emergency Management Assistance Compact
EOC	Emergency Operations Center
ESF	Emergency Support Function
FCO	Federal Coordinating Officer
FEMA	Federal Emergency Management Agency
FRC	Federal Resource Coordinator
HSOC	Homeland Security Operations Center
ICS	Incident Command System
JFC	Joint Force Commander
JFHQ--CA	Joint Force Headquarters--California
JFLCC	Joint Forces Land Component Commander
JFO	Joint Field Office
JIOC	Joint Interagency Operations Center
JOC	Joint Operations Center
JTF	Joint Task Force
LAPD	Los Angeles Police Department
LNO	Liaison Officer

Lt Col	Lieutenant Colonel (Air Force)
LTG	Lieutenant General
NIMS	*National Incident Management System*
NRP	*National Response Plan*
OES	Office of Emergency Services
OPORD	Operations Order
PFO	Principal Federal Officer
POMSO	Plans, Operations, and Military Support Officer
REOC	Regional Emergency Operations Center
SCO	State Coordinating Officer
SEMS	State Emergency Management System
SOC	State Operations Center
TAG	The Adjutant General
TF	Task Force
USACE	US Army Corps of Engineers
USC	*United States Code*
USNORTHCOM	United States Northern Command

ILLUSTRATIONS

TABLES

CHAPTER 1

INTRODUCTION

> We are very resource-rich and also have a very
> sophisticated and long-tested mutual aid system. . . . We learned
> lessons from events dating back to the 1906 earthquake in San
> Francisco. So we have resources in place that we depend on to
> arrive quickly on the scene because they are from our own
> neighbors—cities and counties. . . .We would not depend on the
> Federal Emergency Management Agency to do that for us. . . .
> We've always known it will take time for them to get here.[1]
>
> Henry R. Renteria

Mr. Renteria's comment was made in an interview that appeared in the *Los Angeles Times* on 12 September 2005. Two weeks earlier the United States witnessed one of the most devastating natural disasters in history. The destruction from hurricane Katrina left an estimated 1,330 people dead, 300,000 homes destroyed or uninhabitable, and over 700,000 people displaced.[2] Most of those left homeless were now utterly dependent on the government for food and shelter, and thousands were stranded on rooftops awaiting rescue. The city of New Orleans was virtually uninhabitable, drowning all communications and paralyzing the infrastructure.[3]

The magnitude of the Katrina disaster caused large amounts of resources to deploy at all levels, to include Department of Defense assets. Five days after the hurricane's landfall the Department of Defense began deploying ground forces into the area. While the bulk of the military support was provided by the National Guard forces of Louisiana and Mississippi in the first five days, most of the Guard response came from outside the affected states. In all, 50,000 National Guard and 20,000 active duty personnel participated in the response.[4] One criticism of the massive military response

1

during hurricane Katrina was the inadequate integration of large numbers of deployed troops from different commands during disaster response operations. The most significant problem caused by this lack of integration was a failure in the military's unity of effort. No one had the total picture of forces on the ground, the forces that were on the way, the missions that had been resourced, and the missions still needing to be completed. These situations often lead to a duplication of effort.[5]

In August of 2001, the Federal Emergency Management Agency (FEMA) held a training session that discussed the three most likely catastrophes to strike the United States, A terrorist attack in New York, a super--strength hurricane hitting New Orleans, and a major earthquake on the San Andreas fault.[6] Because of California's geographic location, its volatile and diverse environments, and the state's population density, federally declared emergencies in Californian are the second highest in the country.[7]

On 17October 1989, the San Francisco Bay area was shaken by an earthquake registering 7.1 on the Richter scale. The quake was the most damaging in the United States in 80 years. 62 people were killed and 14,000 residents required emergency shelter.[8] Five years later, a 6.6 Northridge earthquake struck 20 miles northwest of Los Angeles. The death toll caused by the Northridge quake reached 57, 1,500 persons were hospitalized, and over 10,000 people were injured.[9] The State of California has one of the best systems for emergency response of any state in the Union.[10] In part because of the system, the majority of California's disasters is handled at the state and local government levels. California has a sufficient amount of resources to respond to a crisis, and if additional resources are required, it has the ability to manage the situation until national resources can be requested, gathered, and deployed. When the need for federal support has been required, the resource-rich and long--tested mutual aid system referred to by Mr. Renteria has afforded California the time to coordinate and organize the federal resources required to assist in the response.

But catastrophic disasters are a different matter. State and local resources are usually destroyed or exhausted immediately, and authorities may have difficulty determining or communicating their needs.[11] Federal and National Guard forces have both been deployed in response to disasters in California as far back as the 1906 San Francisco earthquake. During the Loma Prieta earthquake over 800 active duty Army soldiers and 1,050 California Guardsmen in state active duty responded to disaster relief operations.[12] In 1992, during the Los Angeles riots, 10,456 Guardsman and 3,531 active duty U.S. Army's soldiers and Marines were deployed.[13] Again, during the Northridge earthquake in 1994, federal troops, including the Army, Navy and Marines, and National Guard forces were called in response to the disaster. Given California's history of disasters, it is highly probable that California will see another catastrophic disaster where the state's ability to provide the necessary emergency response resources is exceeded and federal assistance is required.

The purpose of this study is to determine the most effective command relationship between active forces and the National Guard to provide military unity of effort in response to a catastrophic disaster within California. In order to address this issue, the problem must be clarified. After the attacks on 11 September 2001, the Department of Defense revised the Unified Command Plan and in October 2002 established a new combatant command, US Northern Command (USNORTHCOM), to defend the United States and support military assistance to civil authorities.[14] USNORTHCOM exercises

3

coordinating authority of all homeland defense and homeland security activities of all federal forces and National Guard organizations within the United States. To effectively employ the capabilities of a military organization, the Department of Defense has developed policies on command and control relationships. The relationships are important, because they provide the legal framework that governs the type and amount of authority a commander can exercise.[15] The problem in the military's failure to achieve unity of effort, as seen during operations in Katrina, arises from these command relationships. The command authority for federal forces is established by Title 10, *United States Code*. Unless federalized, command authority for National Guard forces is established under Title 32, and command and control remain with the state. The key in this instance is that under federal law National Guard forces remain under the command and control of the governor, but they are funded with federal dollars.[16]

Under California law, by virtue of his office, the governor is the Commander in Chief of the National Guard. The authority to deploy and use California's National Guard in response to a disaster resides with the Governor.[17]

The question is what command relationship should be established between active and National Guard forces to provide military unity of effort under California's emergency management system?

The first concern is how command relationships are established and what levels of command and control those relationships have over various military units and organizations?

Next, what are the definitions of unity of command, unity of effort, and unified command, and how are they applied to both military and civilian organizations operating under the federal and state emergency management plans and systems?

The third question this study examines is how military support to civil authority operations is coordinated and conducted under the California emergency management system. The *National Response Plan* (2004), *National Incident Command System* (draft 2007), *California State Emergency Plan* (2005), and the Standardized Emergency Management System (2001) all define the emergency management operations and systems used in California. The analysis of these emergency management plans and systems will develop the understanding of the civilian organization and structure under which military forces will operate while conducting military support to civil authority.

Fourth, this study describes, what command relationships and structures can be used to manage military forces in support of civilian authorities. This study will analyze three different models for command relationships between active and National Guard forces during domestic emergency response operations. The first model maintains the separate command structure for active and National Guard forces, maintaining separate command relationships and authorities for Title 10 and National Guard forces in State active duty or Title 32 status operating in the same geographic area. The second model is to federalize all National Guard forces under Title 10 and incorporate them under an active duty command structure. The final model involves the establishment of a "dual-hat" Title 10 Title 32 Headquarters for command and control over all responding military forces, placing both federal Title 10 and National Guard forces in Title 32 status under the control of one chain of command. In answering the question, how to best achieve

5

unity of effort when federal and state forces respond to the same emergency, this study will examine the advantages and disadvantages of each model employed in disaster response.

Finally, the study will examine how active and National Guard command relationships were organized during previous operations. This study will apply the observations and lessons learned about these relationships under the emergency management systems and organizations within California and will compare and contrast them to determine their effectiveness in obtaining military unity of effort.

Assumptions

This study assumes that USNORTHCOM will maintain its current command relationships and authorities over active Title 10 and Title 32 National Guard forces. USNORTHCOM exercises coordinating authority as mentioned earlier. The commander has no authority to direct a unit's action unless or until such forces are placed under his operational control (OPCON).[18] This study further assumes that all forces employed will follow joint doctrine in the establishment and execution of command relationships. The relationships between and among force elements follow a set of principles to establish a chain of command, to facilitate the best possible utilization of all available capabilities, and to ensure unified action in mission accomplishment. It also assumes that Department of Defense resources are used judiciously and adhere to the principles and priorities determined by the President and Secretary of Defense and that the use of military forces under the provisions of the *Posse Comitatus Act*, the Insurrection Act, and Department of Defense Directives are relevant and applicable to the operation.

Definition of Terms

This study deals with military organizations, policies, and doctrine as well as federal, state, and local government and emergency response organizations and management systems. Frequently the common use terms of the military are used differently or carry a different meaning from the same terms in the civilian community. Likewise, the terms used in emergency management and response have specific definitions. In order to avoid confusion and allow a clearer understanding of this study there are certain key terms and words used throughout this document that require defining.

Active and National Guard force

Active forces and active duty refer to those military units and personnel organized under the Department of Defense in the full-time active military service of the United States. This does not include full-time National Guard duty.[19]

National Guard forces refers to the state-organized units of the United States Army and Air Force, composed of citizens who undergo training and are available for service in national or local emergencies. National Guard units are organized in each of the fifty states, the District of Columbia, the Virgin Islands, Guam, and Puerto Rico. National Guard units are subject to the call of the governor of their state or territory, except when ordered into federal service by the president of the United States.[20]

Title 10 and Title 32

Title 10 refers to the portion of the *United States Code* that establishes the Armed Forces of the United States; including the Army, Navy and Marine Corps, Air Force, and

Reserve Components. Title 10 forces are under the command and control of the president of the United States.

Title 32 refers to the portion of the *United States Code* that establishes the organization, personnel, and training of the National Guard. Title 32 status refers to the duty status of National Guard forces by the state in response to requests for assistance. Title 32 forces remain under the command and control of the governor and the state's adjutant general as with state active duty. Funding associated with Title 32 support is furnished by the federal government.[21]

State Active Duty

The governor may call into active service any portion of the state's National Guard as may be necessary in response to public calamity or catastrophe state active duty forces remain under the command and control of the governor and the state's Adjutant General. Funding associated with state active duty is normally furnished by the specific state or territory.[22]

Command

Command can have distinct meanings to both civilians and the military. In the civilian environment, command is the act of directing, ordering, or controlling by virtue of explicit statutory, regulatory, or delegated authority.[23] To the military, command is the authority that a commander in the armed forces lawfully exercises over subordinates by virtue of rank or assignment. Command includes the authority and responsibility for effectively using available resources and for planning the employment of, organizing, directing, coordinating, and controlling military forces for the accomplishment of assigned missions. It also refers to a unit or units, an organization, or an area under the command of one individual.[24]

Combatant Command (Command Authority)

Nontransferable command authority established by Title 10 ("Armed Forces"), *United States Code*, section 164, exercised only by commanders of unified or specified combatant commands unless otherwise directed by the President or the Secretary of Defense. Combatant command (command authority) cannot be delegated and is the authority of a combatant commander to perform those functions of command over assigned forces involving organizing and employing commands and forces, assigning tasks, designating objectives, and giving authoritative direction over all aspects of military operations, joint training, and logistics necessary to accomplish the missions assigned to the command. Combatant command (command authority) should be exercised through the commanders of subordinate organizations. Normally this authority is exercised through subordinate joint force commanders and service and or functional component commanders. Combatant command (command authority) provides full authority to organize and employ commands and forces as the combatant commander considers necessary to accomplish assigned missions. Operational control is inherent in combatant command (command authority).[25]

Limitations

The research in the study is limited by the availability of after--action reviews and reports available from the State of California through electronic means. The restrictions of time and distance limit the ability to review documents prepared by the California

National Guard and the Office of Emergency Services that are maintained by the organizations in hard--copy format. To attempt to avoid the application or perception of possible bias, the research is limited to the published documentation dealing with the emergency response to the analyzed incidents and disasters.

Scope and Delimitations

This study will review the federal and state laws, policies, procedures, organizations, and systems that apply to the use of active and National Guard forces in the execution of Defense Support to Civil Authorities (DSCA) during a natural or man--made disaster that occurs within the State of California. The study does not cover the application or use of active or National Guard forces in the conduct of counterdrug operations or acts of terrorism. It also does not cover the shortfalls or deficiencies in the state or federal policies, plans, or execution of emergency response by the civilian agencies involved in the emergency response process of the analyzed incidents and disasters.

Significance of the Study

Utilizing the results and analysis of this study will identify the areas, coordination, and structure required in the development of future plans and procedures to effectively establish and execute the military's response to a catastrophic disaster within California. The information provided will enable military officials from both the active military forces and the National Guard to provide civilian authorities with a unity of effort in the application of the Department of Defense's capabilities and resources in response to catastrophic disaster. The study further serves as a guide on how to apply the recommendations for the coordination and command relationships between active and National Guard forces responding to disasters both at the national and state level.

The United States has seen its share of catastrophic disasters, and in its response the government has utilized the resources of the Department of Defense to aid the state and local governments to protect the public safety of the population by providing military support to civil authorities. The military is called upon during domestic emergency because it can provide an organized pool of labor and equipment to assist local authorities in saving lives and restoring order in communities devastated by a disaster. California too has relied upon active forces in response to the catastrophes that have affected the state. With the state's National Guard available to the governor and often deployed as part of the state's response capabilities, civilian and military authorities struggle to obtain a unity of effort within the military capabilities and resources supporting emergency operations. The laws and policies that establish the command relationships between active and National Guard forces have complicated and sometimes hindered achieving the required unity of effort. The research in this study attempts to provide clarification for the laws, policies, and procedures involved in utilizing both active and National Guard forces in response to disasters within California. The review of the literature used in this study provides an understanding of those policies and procedures of the emergency management systems and plans that provide for both state and federal response and the use of military forces during domestic disasters.

---

[1]Tim Reiterman, "Q&A/Henry Renteria: Making Sure State is Ready in Case Disaster Strikes" *Los Angeles Times,* 12 September 2005, B2.

[2]United States Government, Department of Homeland Security, *The Federal Response to Hurricane Katrina: Lessons Learned* (Washington, DC: US Government Printing Office, 2006), 7-8.

[3]Christopher Cooper and Robert Block, *Disaster: Hurricane Katrina and the Failure of Homeland Security,* 1std ed. (New York, New York: Time Books, Henry Holt, and Company, 2006), 4.

[4]United States Government Accounting Office, *Hurricane Katrina: Better Plans and Exercises Need to Guide the Military's Response to Catastrophic Natural Disasters* (Washington, DC: US Government Printing Office, 2006), 6.

[5]Ibid., 9.

[6]Jia-Rui Chong and Hector Becerra, "Katrina's Aftermath; California Earthquake could be the Next Katrina," *Los Angeles Times,* 8 September 2005, A.1.

[7]"Federal Emergency Management Agency," 18 May 2007. Accessed 24 May 2007, Available from http://www.fema.gov/news/disaster_totals_annual.fema.

[8]Janet A. McDonnell, *Response to the Loma Prieta Earthquake* (Fort Belvoir, VA: US Government Printing Office, 1993), 1.

[9]Sergey Khomchenko, "Civil-Military Relations in Domestic Support Operations. the California National Guard in Los Angeles 1992 Riots and Northridge Earthquake of 1994" (Ph.D. diss., Naval Postgraduate School, 1997), 57.

[10]Michael A. Wermuth, *Enhancing Emergency Preparedness in California: Testemony Presented to the Little Hoover Commission January 26, 2006* (Santa Monica, CA: RAND Corporation, 2006), 4.

[11]Alane Kochems, "Military Support to Civilian Authorities: An Assessment of the Response to Hurricane Katrina," *Backgrounder*, No. 1899 (28 November 2005 2005): 1.

[12]McDonnell, "Response to the Loma Prieta Earthquake," 9-10.

[13]James D. Delk, *Fires and Furies: The Los Angeles Riots of 1992* (Palm Springs, CA: ETC Publications, 1995), 339-340.

[14]"Unified Command Plan," 3 May 2007 Accessed 24 May 2007; available from http://www.defenselink.mil/specials/unifiedcommand/.

[15]Lynn E. Davis et al., *Army Forces for Homeland Security* (Santa Monica, CA: RAND Corporation, 2004), 69-70.

[16]LTG H. Steven Blum, *National Guard Homeland Defense Whitepaper: September 11, 2001, Hurricane Katrina, and Beyond* (Washington DC: National Guard Bureau, 2005), 14.

[17]"Official California Legislative Information, California Codes Military and Veterans Code," 1 January 2007, accessed 24 May 2007; available from http://leginfo.public.ca.gov.

[18]Davis et al., "Army Forces for Homeland Security," 70.

[19]United States, Department of Defense, Joint Publication 1-02, *Department of Defense Dictionary of Military and Associated Terms* (Washington, DC: Directorate for Operational Plans and Joint Force Development, 2001), 4.

[20]"West's Encyclopedia of American Law, 'National Guard'", accessed 12 May 2007; available from http://www.answers.com/topic/united-states-national-guard.

[21]Cornell University, Law School, "U.S. Code Collection," 30 May 2006, accessed 24 May 2007; Available from http://www4.law.cornell.edu/uscode/.

[22]State of California, Legislative Counsel, "Official California Legislative Information, California Codes Military and Veterans Code,"

[23]United States Government, Department of Homeland Security, *National Incident Management System (Draft),* (Washington DC: U.S. Government Printing Office, 2007), 185.

[24]United States, Department of Defense, "Joint Publication 1-02, *Department of Defense Dictionary of Military and Associated Terms*," 101.

[25]Ibid., 96.

CHAPTER 2

LITERATURE REVIEW

After the devastation that occurred during hurricane Katrina, both the government and the media published numerous accounts of national response in an effort to either fix blame or to identify shortfalls in the system that required improvement. A major issue with the Department of Defense's response was that coordination required for unity of effort proved difficult, and military unity of effort was never achieved. The National Guard troops were placed under the operational control of the adjutant generals of Louisiana and Mississippi, while the active forces fell under the command of USNORTHCOM's Joint Task Force Katrina. These command relationships essentially set up three separate military chains of command.[1] In 1994, after the Northridge earthquake in Los Angeles, California, the California National Guard responded in support of the civil authorities in a State active duty status. Active federal forces were also called upon when the President declared a major emergency in Los Angeles and Ventura counties.[2] In the military response to the Northridge earthquake, there were also separate command relationships for active and National Guard forces. In an effort to improve the methods in which the military provides support to civil authorities during a disaster, this study will determine what command relationship between active and National Guard forces must be established to provide a military unity of effort under California's emergency management system.

This chapter is organized to answer the secondary and tertiary questions proposed in chapter 1 and is prepared in a logical sequence to facilitate the analysis of the literature reviewed to the subsequent material.

Command Relationships

Command relationships are a key element in controlling actions to provide authorities unity of effort amongst the agencies and organizations responding to emergencies and disasters. The review of literature on command relationships provides an understanding of how command relationships function in the military and the civilian organizations responsible for coordinating and executing the response activities during a disaster.

Military Command Relationships

Command relationships are a crucial element of the way the military employs its capabilities and exercises control over its organization and units. Command relationships provide the legal framework that governs the type and amount of authority that commanders may exercise.

The RAND Corporation published *Army Forces for Homeland Security* in 2004 providing an objective and recommendations to the challenges facing the military in the performance of its responsibilities for homeland security. Appendix C provides a simple description of the military command relationships. There are four general types of command relationships in the Department of Defense: combatant command (command authority), coordinating authority, operational control, and tactical control. Combatant command (command authority) is established by law and is exercised only by commanders of unified or specified combatant commands. Combatant command (command authority) cannot be delegated and is the authority of a combatant commander to perform those functions of command over assigned forces. Coordinating authority is the assigned authority of a commander or individual for coordinating specific functions or activities involving two or more military forces, military departments, or joint force components. The commander has the authority to require consultation between agencies, but does not have the authority to compel agreement. Coordinating authority is not an authority to exercise command; it is more applicable to planning than to operations. Operational control is the command authority that may be exercised at any echelon at or below the combatant command. Operational control is inherent in the combatant command and may be delegated within the command. Operational control is the authority to perform those functions of the command over subordinate forces necessary to accomplish assigned missions; it does not include authoritative direction of logistics or matters of administration, discipline, internal organization, or unit training. During hurricane Katrina USNORTHCOM was given operational control over the 82nd Airborne. The Adjutant Generals of Louisiana and Mississippi were given operational control over the National Guard forces responding to the disaster in their areas. Command authority still resided with the providing states but the Louisiana and Mississippi Adjutant Generals were able to direct activities and operations of the National Guard forces within their states. The last command relationship is tactical control. Tactical control is the authority over assigned or attached forces made available for tasking that is limited to the detail of the movement or maneuver within the operational area necessary to accomplish the assigned mission.[3]

Events and political administration have changed the way both the civilian and military organizations and command structures meet the needs of the nation in response to disasters and the treat of terrorism. *Disaster Response and Homeland Security; What works, What doesn't* (Miskel, 2006) points out the despite these changes the basic principles for authorizing military support and coordinating requests during a disaster remains the same. During an emergency, a defense coordination officer (DCO) is assigned and works directly with the federal coordinating officer (FCO), who aids the local authorities in requesting and coordinating federal resources. The defense coordination officer consults with the federal coordinating officer about military options and, if a military support option is selected, the defense coordination officer forwards the request through the appropriate military chain of command. In disasters requiring an extensive military response, the military may be organized under a Joint Task Force (JTF). Under this organization, the defense coordinator role changes to a liaison between the JTF commander and the federal coordinating officer and local authorities.[4]

There are several publications in military doctrine that expand on the brief description of command relationships provided by the RAND Corporation. JP 3-0, *Joint Operations*,(2006) Provides military guidance for the exercise of authority by combatant commanders and other joint force commanders (JFCs) and prescribes doctrine for operations. Besides defining the types and levels of command relationships, JP 3-0 describes command relationships as key considerations to joint functions. Command and control is how commanders exercise their command authority including communicating and maintaining information, assessing the situation, coordinating and controlling the employment of capabilities, and coordinating and integrating support. Control is inherent in command. Control of forces helps the commander determine the requirements, allocate the means, and integrate efforts. It also explains the concept of functional components and the establishment of functional component commands to conduct operations when forces from two or more services must operate in the same domain or there is a need to accomplish a distinct aspect of the assigned mission.[5] An example of a functional component command relationship is the Joint Forces Land Component Commander (JFLCC). In an operation involving both Army and Marine ground forces the JFLCC manages the operations of land forces and reduces the Joint Force Commander's requirement to oversee and influence every task. Joint Publications JP 1, *Joint Warfare of the Armed Forces of the United States* (2000) describes the concept of unified action for the purpose of achieving unity of effort in mission accomplishment. It delineates how command is central in unity of action and that unified command is fundamental to the unity of effort. The command relationships between forces follow a set of principles to establish a chain of command, facilitate the best possible utilization of all available capabilities, and ensure unified action in mission accomplishment. JP 1 identifies two additional command relationships. Supported and supporting relationships between commands facilitate unified action in planning and conducting operations. Support is a command authority established by a superior commander between subordinate commanders when an organization should aid, protect, complement, or sustain another force. In fulfilling his responsibility, the supported commander must coordinate, synchronize, and integrate the activities of the supporting commands.[6] JP 1 outlines the fundamentals of joint doctrine for military participation in interagency operations, such as support and assistance to civil authorities.

The interagency process facilitates unified action by military and nonmilitary participants conducting operations. Interagency organizations that require military participation but are primarily nonmilitary are organized under appropriate lead agencies. Military assistance in domestic emergencies within the United States may be approved by the President or the Secretary of Defense. In these domestic situations, the Constitution, law, and other government directives limit the scope and nature of military actions. The National Guard under the control of the state or territorial government provides a wide variety of support to civil authorities, whereas forces under federal control must adhere to the provisions of the *Posse Comitatus Act* and other related Department of Defense regulations that prohibit or limit active forces from participating in civilian law enforcement activities.[7] JP 1 was published in November of 2000 and does not include recent updates in the *Unified Command Plan* establishing USNORTHCOM and its missions or command relationships. Nor does it include the relationships and responsibilities of the Department of Homeland Security in domestic emergencies. While there is a gap between doctrine and reality, the fundamental concepts of interagency operations during domestic emergencies are still relevant. FM 6-0, *Mission Command: Command and Control of Army Forces*, identifies the authority of command established by the Constitution, and designates the President as the Commander-in-Chief over the Armed Forces. It also identifies public laws, such as the Uniform Code of Military Justices which grants further authority, and responsibility, to require accountability from commanders. Most significant is the description of unity of effort as a principle of command.

Military Command Authority by Law

The laws establishing command authority for the active forces are found under the provisions of Title 10 of the *United States Code*, those applying to the National Guard are provided under Title 32 of the *United States Code*. Combatant command authority applies to active forces The command authority of a combatant commander provides him

19

the authority to give direction to subordinate commands and forces necessary to carry out the missions assigned to the command, including authoritative direction over all aspects of military operations, joint training, employing forces within that command as he considers fit and exercising his authority with respect to selecting subordinate commanders, selecting combatant command staff, suspending subordinates, and convening courts-martial.[8] Command authority of the National Guard is exercised by the Governor of the state through the Adjutant General. "There shall be an adjutant general in each State and Territory, Puerto Rico, and the District of Columbia. He shall perform the duties prescribed by the laws of that jurisdiction."[9] "To secure a force the units of which when combined will form complete higher tactical units, the President may designate the units of the National Guard, by branch of the Army or organization of the Air Force, to be maintained in each State and Territory, Puerto Rico, and the District of Columbia. However, no change in the branch, organization, or allotment of a unit located entirely within a State may be made without the approval of its governor".[10]

Civilian Command Relationships

Under the *National Incident Management System* (2004) (NIMS) command relationships are based on three key organizations of the command and management component of the management system. The Incident Command System (ICS) was developed after the extensive forest fires in California during the 1970s as an effort to correct the lack of integration among firefighting organizations responding to the incident. ICS is a disaster management tool based on a series of rational bureaucratic principles similar to those often discussed in organizational studies as classical management theory. The literature reviewed on this subject examines the functions and organization of the ICS and multi-agency coordination systems to facilitate unity of effort.

In their *Introduction to Emergency Management, Second edition* Haddow and Bullock, provide a comprehensive description of the background, components, and systems involved in the management of disasters. The main function of ICS is to establish a set of planning and management systems that help the agencies responding to a disaster work together in a coordinated and systematic approach. Command is one of the five major management systems within ICS. Under ICS, the command section includes developing, directing, and maintaining communications and collaboration with the multiple agencies on site working with local officials. ICS uses the Unified Command concept as a process that all participating agencies can use to improve overall management. The concept of Unified Command in the ICS allows the integration of multiple government and non-government agencies under one overall response management structure. Unified Command provides a response system that allows various agencies to work together.[11]

After Hurricane Katrina in 2005, the Department of Homeland Security began coordinating a comprehensive review of the National Response Plan (NRP) and National Incident Management System (NIMS). The review began in October of 2006 and is scheduled to be completed by June of 2007. In February, 2007 version 1 of the draft was released to facilitate the review process with stakeholders at all levels. The review in this study incorporates both the 2004 published version and the February 1, 2007 draft. The *National Incident Management System* provides a consistent national model that enables federal, state, local, and tribal governments as well as private and nongovernmental

21

organizations to work together effectively and efficiently to prepare for, prevent, respond to, and recover from incidents, regardless of cause, size, or complexity, including acts of catastrophic terrorism.

The NIMS provides a set of standardized organizational structures that improve integration and connectivity among jurisdictions and represents a core set of doctrine, concepts, principles, terminology, and organizational processes that enables effective, efficient, and collaborative incident management at all levels.[12] Under the concepts and principles of the NIMS and ICS most incidents are managed locally. Incident response can begin with a single responder within a single jurisdiction and rapidly expand to involve multiple responders and multiple jurisdictions requiring significant additional resources and operational support. The command function is clearly established at the beginning of the incident. The agency with primary jurisdictional authority designates the individual on scene responsible for establishing command. Command under ICS encompasses the Incident Commander and their Command Staff. This may be executed as a single Incident Commander or under a Unified Command. An Area Command may also be established with multiple site-specific Incident Commanders or Unified Commands. When a single Incident Commander is designated he will develop incident objectives on which a subsequent incident action plan is based. The Incident Commander approves all requests pertaining to the ordering and releasing of incident resources.[13] The principle of Unified Command allows agencies with different legal, geographic, and functional authorities and responsibilities to work together effectively without affecting individual agency authority, responsibility, or accountability in incidents involving multiple jurisdictions and or multi-agencies.

Unity of Command, Unity of Effort, and Unified Command

The *National Response Plan* (2004) clarifies there are distinct differences in the concepts of command and Unity of Command between the military and those of civilian authorities under the emergency management system. Command and Unity of Command have distinct legal and cultural meanings for military forces and operations. For military forces, command runs from the President to the Secretary of Defense to the Commander of the combatant command to the commander of the forces. (In the case of the National Guard under Title 32 this command runs from the Governor through the Adjutant General to the commander of National Guard forces) The Unified Command concept utilized by civil authorities is distinct from the military chain of command. Nothing in the NRP impairs or otherwise affects the authority of the Secretary of Defense over the DOD, including the chain of command for military forces or military command and control procedures. The Secretary of Defense (or the state's Governor) shall retain command of military forces providing civil support.[14] The NRP defines a Unified Command as; an application of ICS used when there is more than one agency with incident jurisdiction or when incidents cross political jurisdictions. Agencies work together through the designated members of the Unified Command to establish their designated Incident Commanders at a single Incident Command Post and to establish a common set of objectives and strategies and a single Incident Action Plan. The incorporation and functions of a Unified Command during an incident is further clarified in the NIMS.

The *National Incident Management System* (2007) provides an understanding of these command concepts in the civilian emergency management organizations and systems. Three of the management characteristics that contribute to the strength and efficiency of the Incident Command System are: Chain of Command and Unity of Command, and Unified Command. Chain of Command refers to the orderly line of authority within the ranks of the incident management organization. Unity of Command means that every individual has a designated supervisor to whom they report at the scene of the incident. Unity of Command also establishes a command relationship where all members of the unit, team, or group fall under the authority of one responsible commander. These principles clarify reporting relationships and eliminate the confusion caused by multiple, conflicting directives.[15] Unified Command is an important element in multi--jurisdictional or multi--agency incident management. It provides guidelines to enable agencies with different legal, geographical, and functional responsibilities to coordinate, plan, and interact effectively. A Unified Command allows agencies to jointly provide management direction to an incident through a common set of incident objectives and strategies established at the command level. Each agency maintains its authority, responsibility, and accountability. The Unified Command functions as a single integrated management organization. Under a Unified Command, a single individual, the Operations Section Chief, directs the tactical implementation of the Incident Action Plan.[16]

Table 1.   Difference between a Single Incident Commander and a Unified Command

PRIMARY DIFFERENCES BETWEEN A SINGLE INCIDENT COMMANDER AND UNIFIED COMMAND	
Single Incident Commander	Unified Command
The IC is solely responsible (within the confines of his or her authority) for establishing incident objectives and strategies. The IC is directly responsible for ensuring that all functional area activities are directed toward accomplishment of the strategy.	The individuals designated by their jurisdictional authorities (or by departments within a single jurisdiction) must jointly determine objectives, strategies, plans, resource allocations, and priorities and work together to execute integrated incident operations and maximize the use of assigned resources.

Source: United States Government, Department of Homeland Security, "National Incident Management System (Draft)," 55.

As identified previously, the military has different meanings and understandings of Unity of Command and Unified Command. Both Joint Publication JP 1-02 *Department of Defense Dictionary of Military and Associated Terms* (2001) and JP 3-0 *Joint Operations* (2006) have definitions for these terms. To the military, the purpose of Unity of Command is to ensure Unity of effort under one responsible commander for every objective. Unity of Command means that all forces operate under a single commander with the requisite authority to direct all forces employed in pursuit of a common purpose.[17] For the military, a Unified Command is a command with a broad continuing mission under a single commander and composed of significant assigned components of two or more Military Departments that are established and so designated by the President, through the Secretary of Defense with the advice and assistance of the Chairman of the Joint Chiefs of Staff.[18]

The term Unity of Effort, however, is not specifically defined in the NRP or the National Incident Command System, while the term and concept are used in describing the purpose of a Unified Command. The *Department of Defense Dictionary of Military and Associated Terms* does not have a definition for Unity of Effort either, but a clear understanding of the concept can be determined from military doctrine. Army Field Manual FM 6-0 describes the concept of Unity of effort as: the coordination and cooperation among all military forces and other organizations toward a commonly recognized objective, even if the forces and nonmilitary organizations are not part of the same command structure.[19] This concept is further developed in JP 3-0 *Joint Operations* (2006) describing Unity of effort as coordination through cooperation and common interests and is an essential complement to unity of command.[20] From the review of both military doctrine and the National Incident Command System, we can deduce a common understanding and concept for unity of effort. For the purpose of this study, unity of effort during disaster response operations is the coordination and cooperation among all agencies and organizations toward a commonly recognized objective.

<u>Emergency Management Plans, Operations and Systems</u>

When a disaster strikes, the first responders to the site are usually the local police, fire, and emergency medical personnel. When the size of the disaster exceeds the capabilities of the local government, the senior official will request assistance from the governor. The governor, through his or her emergency management office, provides assistance with state resources. When state response assets are insufficient, the governor makes a request to the President for a presidential disaster declaration.

The review of the literature for the processes in emergency management includes *Introduction to Emergency Management, Second Edition* (Haddow, Bullock 2006). This book gives a comprehensive guide to understanding the background and systems of emergency management within the United States. The foundation of emergency management incorporates the sovereignty of the states and territories. Under the U.S. Constitution, the states are given responsibility for public health and safety within their territory. The federal government becomes involved only after the state governor has requested assistance, or when state agencies are unable to fulfill their basic functions.[21] The system involved in emergency response is built on coordination and cooperation among a significant number of federal, state, and local government agencies.[22] As mentioned previously, local police, fire, and medical personnel are the first responders to the scene of a disaster. In fact, these first responders routinely carry out their duties in a systematic and well-planned course of action. The roles and responsibilities of these agencies and organizations are often detailed in community emergency plans. Procedures are often in place to request additional assistance from nearby communities or higher levels of government. At the state level, this process is managed by the state government office of emergency management. The names of these organizations vary from state to state as does where the office resides within the state government. In California, the Office of Emergency Services (OES) is located in the Office of the Governor. Often the principal resource available to governors in responding to a disaster in the state is the National Guard. The National Guard offers the governor substantial capabilities for responding to disasters including; personnel, communication systems, air and ground transportation, heavy construction and earth moving equipment, mass casualty care, and emergency supplies such as cots, blankets, and medical supplies.[23] Once the governor has

determined that the disaster has overwhelmed the state and local capabilities, the

governor forwards a letter to the President requesting a presidential disaster declaration.

This is the first step towards involving federal officials, agencies, and departments and

resources. The presidential disaster declaration makes available the resources of the

federal government. Although a formal declaration does not have to be signed for the

federal government to respond, the governor must make a formal request for assistance

and specify in the request the specific needs of the disaster area. While this sounds like a

time consuming bureaucratic process, situations, threats, incidents, and potential

incidents are continually monitored and reported from federal, state, local, tribal, and

nongovernmental organizations to the Homeland Security Operations Center (HSOC).

The HSOC makes an initial determination regarding these reports, to initiate the

coordination of federal information-sharing and incident management activities. The

Department of Homeland Security, through FEMA, is responsible for coordinating all

federal activities in support of the state and local response and recovery efforts in a

presidential declared disaster. In this instance, FEMA activates the National Response

Plan (NRP).[24]

The National Response Plan

In February 2003, President Bush signed Presidential Directive 5 "to enhance the

ability of the United States to manage domestic incidents by establishing a single,

comprehensive national incident management system"[25] This action authorized the

design and development of the *National Incident Management System* (2004) and the

National Response Plan (2004) replaced the Federal Response Plan written in 1992 and

28

was designed according to the template of the National Incident Management System (NIMS). The NRP provides the framework for federal interaction with state, local, and tribal governments; the private sector; and nongovernmental organizations in the context of domestic incident prevention, preparedness, response, and recovery activities. Within this framework, it establishes a mechanism to facilitate emergency mutual aid and federal emergency support to state, local, and tribal governments.[26]

There are 32 federal department and agencies that are signatories to the NRP. Each of these serves as a primary or support agency in one or more of the plan's Emergency Support Functions (ESF). A primary agency is the federal agency that serves as the federal executive agent under the Federal Coordination Officer to accomplish the ESF mission. Among the responsibilities of the primary agency is to orchestrate federal support within there functional area for an affected state. Support agencies conduct operations when requested, using their own authorities, capabilities, and resources. The ESFs serve as the coordination mechanism to provide assistance to the supported state, local, or tribal governments or federal agencies conducting operations for disaster assistance (figure 2). The Department of Defense is a support agency to 14 of the EFS with its U.S. Army Corps of Engineers the primary agency for ESF #3 (public works and engineering). The purpose of ESF #3 is to facilitate the delivery of services, technical assistance, engineer expertise, and construction management.[27]

Table 2.   Emergency Service Functions

ESF 1 - TRANSPORTATION	ESF 9 - URBAN SEARCH AND RESCUE
Federal and civil transportation support	Life-saving assistance
Transportation safety	Urban search and rescue
Restoration/recovery of transportation infrastructure	**ESF 10 - OIL AND HAZARDOUS MATERIALS RESPONSE**
Movement restrictions	
Damage and impact assessment	Oil and hazardous materials (chemical, biological, radiological, ect.) response
ESF 2 - COMMUNICATIONS	
Coordination with telecommunications industry	Environmental safety and short- and long-term cleanup
Restoration/repair of telecommunications infrastructure	**ESF 11- AGRICULTURE AND NATURAL RESOURCES**
Protection, restoration, and sustainment of national cyber and information technology resources	Nutrition assistance
	Animal and plant disease/pet response
ESF 3 - PUBLIC WORKS AND ENGINEERING	Food safety and security
Infrastructure protection and emergency repair	Natural and cultural resources and historic properties protection and restoration
Infrastructure resoration	
Engineering services, construction management	**ESF 12- ENERGY**
Critical infrastructure liaison	Energy infrastructure assessment, repair, and restoration
ESF 4 - FIREFIGHTING	Energy industry utilities coordination
Firefighting activities on Federal lands	Energy forcast
Resource support to rural and urban fierfighting operations	**ESF 13 - PUBLIC SAFETY AND SECURITY**
	Facility and resource security
ESF 5 - EMERGENCY MANAGEMENT	Security planning and technical and resource assistance
Coordination of incident management efforts	Public safety/security support
Issuance of mission assignments	Support to access, traffic, and crowd control
Resource and human capital	**ESF 14 - LONG-TERM COMMUNITY RECOVERY AND MITIGATION**
Incident action planning	
Financial management	Social and economic community impact assessment
ESF 6 - MASS CARE, HOUSING, AND HUMAN SERVICES	Long-term community recovery assistance to States, local governments, and the private sector
Mass Care	Mitigation analysis and program implementation
Disaster housing	**ESF 15 - EXTERNAL AFFAIRS**
Human services	Emergency public information and protective action guidance
ESF 7- RESOURCE SUPPORT	Media and community relations
Resource support (facility space, office equipment and supplies, contracting services, ect.)	Congressional and international affairs
ESF 8- PUBLIC HEALTH AND MEDICAL SERVICES	Tribal and insular affairs
Public Health	
Medical	
Mental health services	
Mortuary services	

Source: United States Government, Department of Homeland Security, "National Response Plan," 12.

The roles and responsibility of the state governors under the NRP include; coordinating State resources to address the full spectrum of actions to prevent, prepare for, respond to, and recover from incidents in an all-hazards context to include terrorism, natural disaster, accidents, and other contingencies. The governor is the Commander in

Chief of State military forces (National Guard when in State active duty or Title 32 Status and the authorized State militias).[28]

Within the NRP concept of operations in an incident of national significance, the Secretary of Homeland Security, in coordination with other federal departments and agencies, initiates actions to prevent, prepare for, respond to, and recover from the incident. During this type of incident, the overall coordination of federal incident management activities is executed through the Secretary of Homeland Security. The Secretary utilizes multi-agency structures, at the headquarters, regional and local field offices. In the field the Secretary is represented by the Principal Federal Officer (PFO) and or the Federal Coordinating Officer (FCO) or Federal Resource Coordinator (FRC) as appropriate. Overall Federal support to the incident command structure on the scene is coordinated through the Joint Field Office.[29] The role of the regional coordination structure varies depending on the situation. Larger, more complex incidents may require direct coordination between the Joint Field Office and the national level.

The Department of Defense normally provides Defense Support to Civil Authorities (DSCA) when state, local, and federal resources are overwhelmed, provided that it does not interfere with the Department's military readiness or operations. Requests for DSCA assistance are made to the Office of the Secretary of Defense, Executive Secretariat. If approved, the Secretary of Defense designates the supported combatant commander for the response. The supported combatant commander is the military command authority over all active forces supporting civilian authorities; he determines the level of response requirements and directs a senior military officer to deploy to the incident site. This senior military officer is most often the Defense Coordinating Officer

(DCO) and serves as the Department of Defense's single point of contact in the Joint Field Office. The DCO coordinates and processes all requests for military assistance with the exception of US Army Corps of Engineer (USACE) support, or National Guard forces in state active duty status or Title 32. Based on the magnitude, type of disaster, and anticipated level of resources involved, the supported combatant commander may utilize a Joint Task Force (JTF) to consolidate and manage supporting military activities. The JTF commander exercises operational control over all allocated Department of Defense assets (excluding USACE, and National Guard forces in State active duty status or Title 32). National Guard forces employed in State active duty or Title 32 are providing support to the Governor of the State and are not part of the federal military response efforts.[30]

National Incident Management System

The *National Incident Management System* (NIMS) provides the template for incident management regardless of size, scope, or cause of the event. NIMS ensures a consistent doctrinal framework for the management of incidents at all jurisdictional levels regardless of the incident cause, size, or complexity. The benefits of NIMS are standardized organizational structures, processes, and procedures, standardized planning, training, and exercising, as well as information management system.[31] The NIMS incorporates the functions of the Incident Command System.

Incident Command System

The *Incident Command System* (ICS) is the combination of facilities, equipment, personnel, procedures, and communications operating within a common organizational structure, designed to aid in incident management activities. It is used for a broad spectrum of emergencies, from small to complex incidents, both natural and manmade, including catastrophic acts of terrorism. ICS is used by all levels of government--Federal, State, local, and Tribal. The five major management systems within the ICS include Command, (summarized earlier in this chapter), Operations, Planning, Logistics, and Finance. Military representatives may be present or interact with any or all of the ICS staff sections.

The ICS organizational structure is modular, extending to incorporate all elements necessary for the type, size, scope, and complexity of a given incident. The initial responding IC may determine that it is necessary to delegate functional management to one or more Section Chiefs in order to maintain a manageable span of control.[32] Figure 3 depicts the organizational template for an Operations Section. Expansions of this basic structure will vary according to numerous considerations and operational factors. Branches may be used to serve several purposes, and may be functional, geographic or both depending on the circumstances of the incident.

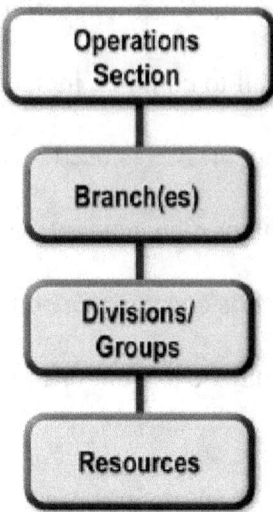

Figure 1.   Major Organizational Elements of the Operations Section
Source: United States Government, Department of Homeland Security, "National
Incident Management System (Draft)," 58.

Divisions and Groups are established when the number of resources exceeds the

manageable span of control of Incident Command and the Operations Section Chief.

Divisions are established to divide an incident into physical and or geographical areas of

operation. Groups are established to divide the incident into functional areas of

operation.[33] Military assets and activities may be organized or coordinate operations

through any of the major elements depending on the scope of the incident and the

Incident Commander desires.

The Incident Commander will establish an Incident Command Post based on the

requirements of the incident and the commander's desires. The Incident Command Post is

located at the local level. Military assets will be represented in the Incident Command

Post by the agency's Liaison Officer (LNO). The LNO is the Incident Command's point

of contact for the military's active duty supported combatant commander or JTF

commander. If the National Guard is employed in State active duty or Title 32, an LNO representing the National Guard forces will be located in the Incident Command Post as well. All military activities are coordinate through the LNO. The military LNO assigned to an incident must have the authority to speak for their organizations on all matters, following appropriate consultations with their leadership.[34]

California's State Emergency Plan and the Standardized
Emergency Management System

The *State Emergency Plan* (2005) defines the emergency management system used for all emergencies in California and describes the California Emergency Organization which provides the Governor access to public and private resources within the State in times of emergency. The State Emergency Plan establishes the policies, concepts, and general protocols for the implementation of Standardized Emergency Management System (SEMS). The use of SEMS is required by law during multi-agency or multi-jurisdictional emergency response by State agencies. The Standardized Emergency Management System is the system for managing response to multi-agency and multi-jurisdiction emergencies in California. SEMS consists of five organizational levels which are activated as necessary: field response, local government, operational area, region, and State (Figure 4). SEMS incorporates the use of the ICS, the Master Mutual Aid Agreement, existing discipline specific mutual aid, the operational area concept, and multi-agency or inter-agency coordination. SEMS helps unify all elements of California's emergency management organization into a single integrated system. Its use is required for State response agencies.

The California Emergency Organization collectively refers to the five organization levels of the SEMS (Figure 5). This organization represents all resources

35

available within the State which may be applied in disaster response and recovery phases. It operates from established Emergency Operations Centers (EOCs) at all levels of government. The goal is to support emergency activities to protect life, property, and the environment. During a state of emergency, or a local emergency, the Office of Emergency Services (OES) Director will coordinate the emergency activities of all State agencies.[35]

There are three OES Administrative Regions (Inland, Coastal, and Southern) in California. The State OES Administrative Regions manage and coordinate information and resources among operational areas and State agencies for support during emergency response, and recovery activities through the Regional Emergency Operations Center (REOC). California is comprised of 58 operational areas. The operational areas consist of all political subdivisions within a county's geographical area. It provides communication and coordination between local jurisdictions and OES Regions. Coordination between the operational area and local government is accomplished through the operational area Emergency Operations Center.[36]

The emergency organization in California are disciplined, organized and structured according to the SEMS model, and relate to the Emergency Service Functions of the NRP. The California Emergency Plan assigns responsibilities to State agencies as lead and supporting agencies similar to the primary and supporting agencies in the National Response Plan.

During response activities, the Operations function under the SEMS is organized into seven branch areas; Fire and Rescue, Law Enforcement, Medical and Health Services, Care and Shelter, Construction and Engineering, Utilities, and Hazardous Material.[37] Military activities, both active and National Guard, may coordinate and support any or all of the operations branches. Requests for military assistance, whether active or National Guard, originate through the Regional Emergency Operations Center (REOC) and are coordinated through an LNO in the same manner the LNO advises and coordinates military activities under the NRP and the NIMS.

Command Relationship Options for Organizing Military Forces

The command relationships and duty statuses of the National Guard are duly described in the article *The Role of the National Guard in National Defense and Homeland Security*.[38] The National Guard can be employed in support of homeland defense and disasters in three legally distinct ways. First is by the Governor for a state purpose which is authorized by the State's law. Under this method of employment, the National Guard is in a state active duty status. In this case, the command relationship of the National Guard is under the command authority and control of the Governor through the Adjutant General of the state. Fiscal responsibility for the use of the National Guard in this status also resides with the state. Next, with the concurrence of the President, the Governor may employ the National Guard for a primarily federal purpose in a Title 32 status. Under Title 32 status the command relationship of the National Guard remains under the command authority and control of the Governor. The recent enactment of Title 32 *United States Code* Chapter 9, provides the legal authority for the use and funding of National Guard for the purpose of homeland defense. The last status that the National Guard can deploy in is when federalized by the President for a federal purpose under a Title 10 duty status. When federalized, the command relationship of the National Guard is under the exclusive command authority and control of the President and federal military officials. While in a Title 10 duty status, the Governor has no authority over the National Guard, even when they are conducting operations within the state.[39]

The *Joint Forces Quarterly* article *The National Guard and Homeland Defense* offers an explanation of how and why each of these various duty statuses is used.[40] The National Guard holds this dual status based on the concept that we are a union of sovereign states. In the foundations of our country the founding fathers understood the importance of the citizen soldier. The militia clause of the Constitution provides the basis for Congress to call up the militia to execute the laws of the Union in the service of the United States, but authorizes the states to appoint officers, and to train the militia. This dual status benefits the states by maintaining their authority and providing them access to organized and equipped forces able to respond to disasters and emergencies.[41] The National Guard, operating under its command relationship with the Governor, can directly help enforce the law. While under state command and control, National Guard forces are not restricted by the *Posse Comitatus Act*, which under most circumstances prohibits federal forces from being used for law enforcement.

The *Cornell University, Law School* provides a web page with the most recent official version of the *United States Code* made available by the U.S. House of Representatives. A review of both Title 10 and Title 32 of the *United States Code* through this site provides an understanding of the law governing the use of military forces, both active and National Guard, for disasters and homeland defense. The specific command model being analyzed by the study is the combining of Title32 and Title 10 forces under the command authority and control of one joint task force commander. This allows the Governor to maintain state control over the state's National Guard forces employed for disaster response in the state. The legal authority for such a status is provided by 32 Unites States Code, sec. 325.[42]

History of Active and National Guard Command Relationships during Operations

The employment of both active and National Guard forces in response to state and national disaster has been documented and examined on several occasions. Among the most recent are those addressing the response and activities surrounding hurricane Katrina.

The United States Government Accountability Office report, *Hurricane Katrina, Better Plans and Exercises, Need to Guide the Military's Response to Catastrophic Natural Disasters* (2006) recognized the military did not plan for the integration of large numbers of both active and National Guard forces and the National Guard and federal response was coordinated over several chains of command.[43] The conclusions in this

report identified several actions required to improve the military's response during

catastrophic disasters. It identified the need for extensive exercises between the

Department of Defense and civil authorities to alleviate and streamline the challenges

brought about by the large number of organizations and levels of government involved.[44]

Additionally the *Joint Center for Operational Analysis, Quarterly Bulletin*,

Volume VIII, Issue 2, "Katrina" (2006) offers an article by Col Greg Gecowets, USAF

Military Analyst, "Coordination, Command, Control, and Communications", where he

focuses on the issues that hindered an effective Unity of effort. With the National Guard

of 54 states and territories under the command and control of the Adjutant Generals of

Louisiana and Mississippi and active assets under the USNORTHCOM JTF Katrina,

numerous helicopter search and rescue crews were employed in an uncoordinated manner

without formal airspace control measures.[45] National Guard forces were organized into

two state-led task forces: JTF Pelican in Louisiana and JTF Magnolia in Mississippi.

Federal active forces were organized under JTF Katrina commanded by LTG Honore.

Despite the establishment of a Joint Operational Area, LTG Honore only had

coordinating authority with the Adjutant Generals and commanders of the State JTFs.

This led to confusion over the roles and responsibilities of both forces.[46] In an effort to

alleviate the confusion caused with these command relationships, a Dual--Hat command

option was proposed give the commander of JTF Katrina authority over both active and

National Guard forces in Title 32. This proposal was rejected by Governor Blanco.

Although other issues were resolved through personal relationships at senior levels on the

ground; the issue of unity of command was never resolved. The author recommends solutions for the best unity of effort through developing improved integration of headquarters, such as Dual--Hat commanders.[47]

Christopher Cooper and Robert Block have also written a book *Disaster: Hurricane Katrina and the Failure of Homeland Security* that reviews the breakdown in the emergency management systems and identifies areas where coordinated and integrated military response would benefit the response process.

The book *Loma Prieta Earthquake* (1993) focuses primarily on the efforts of the US Army Corps of Engineers, it offers an account of the overall response to the 1989 Loma Prieta earthquake that struck and devastated the bay area in northern California. During the response, both active and National Guard forces were mobilized to the disaster. At that time, the U.S. Sixth Army, headquartered in Presidio, led the Army's response. California National Guard units also responded in a state active duty status. The California National Guard assisted primarily in aviation support to law enforcement and damage assessment operations. Federal agencies, including the Army, were hampered because the Federal Response Plan that was in place at the time was not activated. The size of the disaster did not require the activation of all federal agencies called for in the plan. The Sixth Army had difficulty locating and contacting the offices and agencies they needed to work with. City and State official were reluctant to request federal assistance, primarily due to the cost sharing factors involved. An overall coordinated effort that brought to bear the total capabilities of the National Guard and active forces was never integrated. The Sixth Army claimed to have tremendous assets available but they were never requested.[48]

Fires & Furies: The L.A. Riots (1995) recounts the actions and events that

transpired beginning on April 29th, 1992 when a not guilty verdict in the case of four Los

Angeles police officers charged with assaulting Rodney King spawned civil unrest and

riots that lasted six days. When the situation exceeded local law enforcement capabilities,

Mayor Bradley requested assistance of 2,000 National Guard troops from Governor, Pete

Wilson.[49] In all, the California National Guard responded with over 10,000 soldiers. As

requests for more National Guard troops were made, Guard leaders did not convince the

Governor that Guard deployments were on track. Political pressures over the perceived

delayed response of the Guard led to the Mayor and Governor calling for federal troops

and the federalization of the Guard.[50] When the Guard was federalized the command

relationships between the Guard and active forces changes, the Guard's chain of

command now ran through the Joint Task Force--Los Angeles (JTF--LA), commanded by

Major General Covault. Many changes happened in the use of military forces. Prior to

their federalization while in State active duty status, the California National Guard was

able to support the LAPD with enforcing the law and was not subject to the restrictions of

the *Posse Comitatus Act*. Requests for assistance underwent scrutiny to determine

whether the request was a law enforcement or military function. Delk concludes that this

led to the refusal of many of the LAPD requests.[51]

The *Federal Response to Hurricane Katrina Lessons Learned* (2006) identified

that separate command structures between the National Guard and active military forces

hindered Unity of effort.[52] In an effort to eliminate this shortfall, the report recommends

collocating Federal, State, local and National Guard leaders to enhance the unity of effort.

It further recommends the collocating a single Department of Defense point of contact at

43

the Joint Field Office to coordinate military resources. Additionally, it recommends the establishment of a JTF--state to rapidly deploy forward into the affected area that can provide situational awareness and serve as the initial command and control of both state National Guard and USNORTHCOM federal forces. The JTF--state model streamlines the command structure over all assigned forces supporting civil authorities. The JTF--state would assume command and control over all active and National Guard forces. The familiarity of the JTF--state National Guard commander with affected operational areas and the emergency operation systems and organizations within the state would provide both a unity of effort and Unity of Command.[53]

---

[1]J. Emery Midyette Jr., "Resource and Structure of States' National Guard," *Joint Center for Operational Analysis, Quarterly Bulletin* 8, no. 2 (June 2006): 37.

[2]Khomchenko, "Civil-Military Relations in Domestic Support Operations. the California National Guard in Los Angeles 1992 Riots and Northridge Earthquake of 1994," 59.

[3]Davis et al., "Army Forces for Homeland Security," 69-72.

[4]James F. Miskel, *Disaster Response and Homeland Security: What Works, What Doesn't* (Westport, CT: Praeger Security International, 2006), 52-53.

[5]United States, Department of Defense, Joint Publication 3-0, *Doctrine for Joint Operations* (Washington, DC: United States Joint Forces Command, Director for Operations, 2006), sec. II 12, III 1-7.

[6]United States, Department of Defense, Joint Publication 1, *Joint Warfare of the Armed Forces of the United States* (Washington, DC: Directorate for Operational Plans and Joint Force Development, 2000), sec. V, 7-10.

[7]Ibid., sec. VI 1-6.

[8]Cornell University, Law School, "U.S. Code Collection," 30 May 2006, accessed 24 May 2007; available from http://www4.law.cornell.edu/uscode/.

[9]Ibid.

[10]Ibid.

[11]George D. Haddow and Jane A. Bullock, *Introduction to Emergency Management,* 2nd ed. (Burlington, MA: Elservier Butterworth-Heinemann, 2006), 88-89.

[12]United States Government, Department of Homeland Security, "National Incident Management System (Draft)," 3-8.

[13]Ibid., 53.

[14]United States Government, Department of Homeland Security, *National Response Plan,*(Washington, DC: U.S. Government Printing Office, 2004), 10.

[15]United States Government, Department of Homeland Security, "National Incident Management System (Draft)," 50.

[16]Ibid., 53-54.

[17]United States, Department of Defense, Joint Publication 3-0, *Doctrine for Joint Operations,* A-2.

[18]United States, Department of Defense, Joint Publication 1-02, *Department of Defense Dictionary of Military and Associated Terms*, 561.

[19]United States, Department of the Army, Field Manual 6-0, *Mission Command: Command and Control of Army Forces,* (Washington, DC: Headquarters, Department of the Army, 2003), 2-7.

[20]United States, Department of Defense, Joint Publication 3-0, *Doctrine for Joint Operations,* A-2.

[21]Haddow and Bullock, "Introduction to Emergency Management," xi.

[22]Ibid., 78.

[23]Ibid., 84-85.

[24]Ibid, 91.

[25]"Homeland Security Presidential Directive/HSPD-5," in Office of the Press Secretary [database online]. Washington D.C. 28, February 2003 accessed 12 May 2007; available from http://www.whitehouse.gov/news/releases/2003/02/20030228-9.html.

[26]Haddow and Bullock, "Introduction to Emergency Management," 97.

[27]United States Government, Department of Homeland Security, "National Response Plan," ESF # 1-15.

[28]Ibid., 8.

[29]Ibid., 15.

[30]Ibid., 41-42.

[31]Haddow and Bullock, "Introduction to Emergency Management," 96-97.

[32]United States Government, Department of Homeland Security, "National Incident Management System (Draft)," 107.

[33]Ibid., 59.

[34]Ibid., 56.

[35]State of California, Governor's Office of Emergency Services, *State of California Emergency Plan,* (Sacramento, CA: Office of Emergency Services, Planning Section, 2005), 6.

[36]Ibid., 8.

[37]Ibid., 54.

[38]Timothy J. Lowenberg, "The Role of the National Guard in National Defense and Homeland Security," *National Guard* 59, no. 9 (Sep 2005).

[39]Ibid., 99.

[40]Raymond E. Bell Jr., "U.S. Northern Command and the National Guard," *Joint Forces Quarterly* , no. 36 (2005)

[41]Ibid., 36-40.

[42]Cornell University, The Legal Information Institute, "U.S. Code Collection,"

[43]United States Government Accounting Office, "Hurricane Katrina: Better Plans and Exercises Need to Guide the Military's Response to Catastrophic Natural Disasters," 9.

[44]Ibid., 13.

[45]Lt Col Greg Gecowitz, USAF, "Coordination, Command, Control, and Communications," *Joint Center for Operational Analysis, Quarterly Bulletin* VIII, no. 2 (June, 2006 2006): 17.

[46]Ibid., 22.

[47]Ibid., 23.

[48]McDonnell, "Response to the Loma Prieta Earthquake," 10-19, 67-68.

[49]Delk, "Fires and Furies: The Los Angeles Riots of 1992," 37.

[50]Christopher M. Schnaubelt, "Lessons in Command and Control from the Los Angeles Riots," *Parameters, the US Army's Senior Professional Journal* Vol. XXVII, no. Summer, 1997 (2007), 97

[51]Ibid., 305.

[52]United States Government, Department of Homeland Security, "The Federal Response to Hurricane Katrina: Lessons Learned," 55.

[53]Ibid., 95.

CHAPTER 3

RESEARCH METHODOLOGY

The purpose of this study is to determine the most effective command relationship between active forces and the National Guard to provide a military unity of effort in response to a catastrophic disaster within California.

The organization of chapter 3 presents the methodology used to research the purpose of this study. This chapter discusses the application of this methodology and how it is used to present the research and analysis in the following chapter.

Methodology

The analysis of this research starts with examining the definition of unity of effort as it applies within the NIMS and California's SEMS. This examination will provide the concept in which the remainder of the study will base its analysis. The study then applies this concept to the three command relationships outlined in chapter 1 and identifies how each of those command relationships are organized and interacts in the emergency management systems.

The study uses the definitions of unity of command and unified command to establish the pros and cons of each of these command relationships and organizational structures towards accomplishing a unity of effort.

This thesis then examines the plans and emergency management systems utilized during a catastrophic disaster within California. This review will study *California's State Emergency Plan* (2005) to understand the roles and coordination requirements for the various governing, lead, and supporting state agencies and departments responding during disasters. Through the examination of the Standardized Emergency Management System and the Incident Command System, the study gains an understanding of the procedures and policies that facilitate, organize, and provide the incident command and control structure and allow the conduct of operational missions. To answer the primary question this study must include those plans and systems that incorporate the federal--level response. Therefore, it examines the NRP and the NIMS and its integration and incorporation of Department of Defense activities within the emergency management system in California. The NRP identifies how and when federal assets are employed during a disaster while the NIMS outlines the federal systems, organizations and command and control. The knowledge and understanding of these plans and systems provides the answer to the third question and leads to a recommended solution to the problem.

The study examines three models for command relationships and examines the effectiveness of these models in three case studies; The Los Angeles Riots, Hurricane Katrina, and Operation Winter Freeze, an operation supporting US Customs and Border Protection's Border Patrol along a 295-mile stretch of the US-Canadian border to prevented illegal alien access into the United States. The study conducts a descriptive analysis in each of these cases to identify the advantages and disadvantages of the command relationships under each model.

The plans, systems, and laws in place during the time of the incidents are different from each other. Changes in emergency plans and systems are incorporated in the current plans and systems in California. Neither the NRP nor the Dual--Hat command relationship has been executed in California, but each of the cases demonstrates the difficulties and complexities that arise when both Title 10 and 32 forces are deployed.

The majority of the analysis is derived from examining the after--action reviews, individual accounts, GAO reports, and lessons learned from the two major disaster incidents and applying the lessons learned and analysis to the current plans and emergency management systems used in California. The third case study is hypothesized based on the understanding of the current plans, systems, laws, and policies that impact DSCA operations and command relationships.

The methodology presented in this chapter outlines the research and analysis the thesis follows in chapter 4 to answer the secondary questions and address the primary question of what is the most effective command relationship between active forces and the National Guard to provide a military unity of effort in response to a catastrophic disaster within California. It also provides the logical structure to make recommendation for future command relationships during a major domestic disaster within California in chapter 5.

CHAPTER 4

ANALYSIS

From the reviewed literature, this study is able to define Unity of Effort during

disaster response operations as the coordination and cooperation among all military

forces and civilian organizations toward a commonly recognized objective. Through the

principles of war, the United States military accomplishes unity of effort through

organizational structure and command relationships. A problem occurs when active and

National Guard forces are employed depending on the status under which National Guard

forces are employed. This study outlined three possible command relationships for active

and National Guard forces during disaster operations. The purpose of this study is to

determine the most effective command relationship between active and National Guard

forces to provide military unity of effort in response to a catastrophic disaster within

California.

This chapter is organized in a logical sequence, initially analyzing how the

military, both active and National Guard are integrated and activities coordinated under

the emergency management systems used in California. The next portion of this chapter

looks at the organization and command relationships under the three possible command

relationship models and how each of these relationships is integrated into the emergency

management systems. Finally, this study will analyze the advantages and disadvantages

for each command relationship model during DSCA in response to a disaster within

California.

The government, both federal and state, has called upon the military to provide

support to civilian authorities during emergencies and disasters throughout our nation's history. Initially it is the state's National Guard who is called upon by the Governor to provide the initial military support and capabilities. When this happens, the National Guard is typically providing assistance in State active duty status. The National Guard is under the command and control of the Governor and has primary responsibility for providing military support to state and local civilian authorities during a disaster.[1] When conditions exceed the capabilities of the National Guard, additional federal military support may be requested by the Governor.

When a disaster of occurs in the state of California, or in any community within the United States, the initial response is made by local authorities. The management of emergency response relies on the normal authority and responsibility of local government.[2] The local government agencies, typically law enforcement or fire officials, accomplish immediate response actions within the affected area. This includes requests for mutual aid from adjacent local authorities or state resources. Requests for mutual aid originate at the lowest level of government and are progressively forwarded to the next higher level by local and regional authorities until the request is filled. California uses the *Standardized Emergency Management System* (SEMS) for managing the response to multi agency and multi jurisdictional emergencies.[3] The SEMS incorporates the Incident Command System (ICS) and was reviewed by the Office of Emergency Services in September, 2005 to ensure its compliance with the concepts of NIMS. The SEMS follows the same principles of ICS to unify the elements of emergency management at the state level, as NIMS does at the national level, providing a single integrated system to coordinate the levels of California's emergency management organization. The organization represents all available resources within the State which may be employed

52

in response to an emergency or disaster. Each level of the organization operates through established Emergency Operations Centers to support response and recovery activities. When state level assets are involved, the Director of the Governor's Office of Emergency Services (OES), under the authority of the Governor, coordinates the emergency activities of all State agencies. This authority is often delegated by the Director to the State Operations Center (SOC) or the OES Regional Emergency Operations Centers (REOC).[4] When a local area requires state level assistance, local government officials forward requests for assistance to the REOC responsible for the affected area in their region.

Request for DSCA from the California National Guard, as part of the State's resources and in a State active duty status, are made through the REOC or SOC to the State's plans, operations, and military support officer (POMSO) at the California Joint Force Headquarters (JFHQ--CA). The POMSO within each state or territory coordinates plans and activities for disaster response and recovery support missions.[5] The California POMSO operates under the authority of the Adjutant General and validates requests for National Guard support. During an emergency of significant impact or a catastrophic disaster, the Adjutant General, or POMSO acting under the Adjutant General's authority, assigns a unit or establishes a Joint Task Force (JTF) with overall responsibility for tactical level operations in support of civil authorities.

The POMSO allocates the necessary units, organizations, and capabilities to the

JTF Commander and establishes the appropriate command relationships for him to

execute the requested activities. The JTF Commander coordinates directly with the

Incident Commander or Unified Command to provide the necessary resources and

accomplish the goals of the action plan within the intent and parameters of the assigned mission. The JTF Commander or the Joint Operations Center (JOC) at the JFHQ--CA will assign a Liaison Officer (LNO) to the Incident Command Post and each EOC supporting operations. The LNO serves as the link between the EOC, the JTF Commander, and the POMSO for DSCA missions and activities. The POMSO, through the JFHQ--CA JOC, serves as the single point of contact to resource and coordinate DSCA requests for the California National Guard. For example, if the JTF Command's mission is to provide equipment and personnel to evacuate civilians from an area affected by a flood, the JTF Command takes instructions from the Incident Commander on the areas to evacuate and where to transport displaced civilians. If the Incident Command requires support in either supporting the evacuation center or providing security to the evacuated area, the request for assistance is forwarded through the REOC or SOC to the JFHQ--CA JOC and the POMSO for validation. Upon validation and acceptance of the mission the POMSO provides the resources and the command relationship of the units to the JTF Commander to accomplish the mission. Command authority for all National Guard units remains within the state's military chain of command.

When state resources are exhausted or overwhelmed, the Governor may request federal assistance under a presidential disaster or emergency declaration. The Secretary of Homeland Security is the federal authority responsible to the President for coordinating the federal response to emergencies and disasters.[6] The Secretary of Homeland Security utilizes multi-agency structures at the headquarters, regional, and field levels to coordinate efforts and provide appropriate support to the incident command structure. At the field level overall federal support to the incident command system is coordinated through the Joint Field Office (JFO). The JFO serves as the link between the

SOC and the higher federal regional and national-level operational coordination centers and serves as the central location for the coordination of federal, state, local, and other organizations with responsibility for response and support to the incident. The JFO does not manage on-scene operations. The JFO focuses on providing support to on-scene efforts and conducting broader support operations.[7]

When federal active military resources are required, and approved by the President and the Secretary of Defense, the Department of Defense designates a supported Combatant Commander, usually USNORTHCOM, and establishes the necessary forces and supporting organizations for the mission. The Combatant Commander appoints a Defense Coordination Officer who deploys with a Defense Coordinating Element (DCE) to the JFO. The DCO is the single point of contact in the JFO to coordinate all Department of Defense support to the disaster.[8] This does not include coordination of National Guard support in a State active duty or Title 32 status. The responsibilities for the DCO include processing requirements for military support and forwarding missions through the appropriate military channels to the responding unit or organization. The DCO is not in the military chain of command, but processes the requirements for all deployed Department of Defense activities and personnel in support of the disaster, except USACE as part of ESF #3, or National Guard elements in State active duty or Title 32 status.[9] If the Combatant Commander establishes a JTF, the JTF Commander has operational control of all allocated forces. In this case, the DCO will normally work for the JTF Commander, but remains the point of contact for military support requests at the JFO.[10] Just as with National Guard units, command authority over all active military forces remains in the federal military chain of command from the President through the Secretary of Defense.

Coordination of both federal and state resources in response to a disaster is challenging. It requires constant communication and timely situational awareness at all levels of operation. The JFO is established locally to facilitate the coordination of a multi-agency response. The JFO activities are directed by the JFO Coordination Group. The JFO Coordination Group functions as a multi-agency coordination entity and usually includes the Principal Federal Official (PFO), the Federal Coordinating Officer (FCO), the State Coordinating Officer (SCO), and Senior Federal Officials. In accordance with NIMS and the ICS principles the JFO structure includes a coordinating staff. The size and participants on this staff are determined by the type and magnitude of the incident. When federal active military forces are supporting the incident, the DCO is a member of the staff. The State Coordinating Officer manages the incident activities and programs, and those of the Governor's authorized representative, who is empowered to execute all necessary documents for federal assistance on behalf of the state.[11] The activities and missions of the National Guard responding as part of the State's resources would be represented by the SCO in the coordination of response organizations and activities. It is the responsibility of the JFO Coordination Group to resolve any policy issue and allocate resources to National Guard and active forces. The JFO, including the SCO and the DCO, must have a clear understanding of the operational situation to ensure unity of effort throughout the entire military response.

Organizational Command Relationship Models

Separate Command Relationships and Authorities

The command relationship between active and National Guard forces in this model maintains the established separate chains of command for both organizations. The requests for assistance follow separate channels for state and federal resources. Request for support from the National Guard are handled as any request from state agencies. Requests are validated at the State Operations Center (SOC) and forwarded to the JFHQ-CA JOC and the POMSO for approval and to be resourced. Requests for support from active Department of Defense resources are vetted through the DCO at the JFO and forwarded through the established military channels to the Combatant Commander or JTF. The command relationships between active and National Guard forces conducting response operations in the affected area is that each of them operates under its own chain of command. The active forces command authority runs from the President and the Secretary of Defense to the Combatant Commander and the deployed units. The National Guard authority runs separately from the Governor and the Adjutant General to the deployed units. This command relationship is applicable when National Guard forces are in either State active duty or Title 32 status.

Hurricane Katrina Lessons Learned (Separate)

Hurricane Katrina made landfall as a powerful Category 3 storm at 6:10 am CDT on 29 August, 2005 in Plaquemines Parish, Louisiana. Katrina produced a storm surge as high as twenty-seven feet in Louisiana and Mississippi. During the day of the 29th, two levees broke in New Orleans flooding the city and wreaking havoc throughout the area. The destruction included damage to approximately 350,000 homes and the displacing of approximately 200,000 evacuees to shelters.[12] Surge waters flooded over six miles inland

in many parts of coastal Mississippi and up to twelve miles inland along rivers and bays.

The flooding destroyed New Orleans, the Nation's thirty-fifth largest city. Likewise

Mississippi suffered extensive damage. The city of Biloxi was "decimated," according to

municipal government spokesman Vincent Creel. The storm inflicted a terrible toll of

human suffering, killing at least 1,330 and injuring thousands.[13] As Hurricane Katrina

headed toward the coast, the Louisiana National Guard declared a full alert. The

Louisiana leadership stood up a Joint Operations Center (JOC) and the Louisiana

Governor activated the National Guard.[14] Likewise, the Mississippi National Guard

alerted their soldiers, activated their Emergency Operations Centers, and both states sent

liaison officers to the coastal communities and parishes expected to be impacted by the

storm.[15]

After the hurricane made landfall, local and state responders were quickly

overwhelmed. Governors Barbour and Blanco requested additional National Guard assets

from other states through the Emergency Management Assistance Compact (EMAC) to

assist State and local emergency responders.

On August 30, Deputy Secretary of Defense Gordon England authorized

USNORTHCOM to take all appropriate measures to plan and conduct disaster relief

operations in support of FEMA. USNORTHCOM established Joint Task Force Katrina

(JTF--Katrina) to coordinate the growing military response to the disaster and deployed

DCOs to all the potentially affected states.[16] Most deployments began after President

Bush declared a state of emergency on August 30 and an Incident of National

Significance on August 31. The standard National Guard deployment coordination between State Adjutants General (TAGs) was effective during the initial response but was insufficient for such a large-scale and sustained operation. To address this shortfall, LTG Blum, Chief of the National Guard Bureau, held a conference call on 31 August with all fifty-four TAGs to distribute requests for forces and equipment to all TAGs. That same day, The California National Guard Joint Forces Headquarters issued Operations Order (OPORD) 05-31, Gulf Coast Relief. The order organized Joint Task Force-California (JTF--CA) in response to the request from the National Guard Bureau for a 500 person security force with a Command and Control Headquarters, lead by a Colonel for relief actions in the Eastern Louisiana Area of Operations. The OPORD directed the Commander of JTC--CA to report directly to TAG CA.[17] By 1 September 2005, JTF--Katrina, commanded by LTG Honoré, included approximately 3,000 active duty personnel in the disaster area and by 2 September, nearly 22,000 National Guard soldiers and airmen had deployed to the region. As the situation deteriorated, the Department of Defense sent in additional active duty ground forces, including elements of the 82nd Airborne and 1st Cavalry Divisions, which arrived on 5 September. In all, the Defense Department had 42,990 National Guard members, 17,417 active duty personnel, 20 ships, 360 helicopters, and 93 fixed--wing aircraft in the affected area by 7 September.[18] Active duty military and National Guard personnel provided critical emergency response and security support to the Gulf Coast during the height of the crisis. State active duty and Title 32 National Guard forces that deployed to Louisiana and Mississippi operated under the command of their respective Governors. The Title 10 active duty forces, on the other hand, fell under the command of the President and had more limited civil response authority. Also, the Commanding General of JTF-Katrina and the Adjutant Generals

60

(TAGs) of Louisiana and Mississippi had only a coordinating relationship, with no formal command relationship established, resulting in confusion over roles and responsibilities between active and National Guard forces.[19] In his after--action review comments Lt Col Jeffery Richard, California Air National Guard, as Commander of Joint Security Task Force at New Orleans Airport stated; "There never seemed to be one agency that wanted to be in charge of the situation. That changed with the arrival of the 82nd [Airborne] Division. We talked with their leadership about roles and responsibilities, Title 10 vs. Title 32 to try and find common ground. They eventually established a Joint Interagency Operations Center (JIOC)."[20] Colonel Scott Johnson, Commander JTF--CA, expands on the command relationships between National Guard and active forces at the airport. "Initially, the 82nd [Airborne Division] Colonel informed LtCol Richard that his force had arrived to relieve in place any existing military at the Airport. . . . [I]t was determined that [California's] Joint Security Force needed to remain, since under Title 32 CNG personnel could arm their weapons and Title 10 could not. A division of responsibility took place and through most of the mission. . . .[T]he Title 32 and Title 10 personnel performed the Airport security mission together."[21] Once forces arrived in the area, they fell under separate command structures rather than one single command.[22] Besides JTF Katrina, Command and Control Headquarters for all Title 10 active duty forces, there were JTF Headquarters established in both Louisiana and Mississippi. Each of these command and control elements established subordinate Task Forces within different areas of operation. For example, JTF--CA had Operational Control of all California National Guard forces deployed in support of Gulf Coast response efforts in both states. JTF--CA had to coordinate with several Task Force

Headquarters and civilian agencies for mission requests and support operations. During the response to Hurricane Katrina active and National Guard forces had to coordinate over several chains of command.[23]

The examination of the military's response during Hurricane Katrina identifies some advantages to this command relationship model. Both types of forces maintain organic and habitual support and command relationships within their organization. The administrative support requirements for the organization run through the respective commands and headquarters. The National Guard Joint Force Headquarters in each state is responsible for all pay and administrative needs of its deployed forces, although, maintaining this support is more difficult when forces are deployed out of state. In a natural disaster requiring the use of military for law enforcement support, the separate command relationships between Title 10 and Title 32 allows for a more effective allocation of resources. National Guard forces can be employed in support of law enforcement without violating the *Posse Comitatus Act*.

The disadvantages identified with the command relationship during Hurricane Katrina are more extensive, and were a significant contributor to the lack of unity of effort. There was no unity of command binding all military forces supporting the operation. Requests for military assets and assistance were run over several different chains of command. Additionally, the National Guard Bureau has no command or tasking authority over National Guard forces in any state. Requests for additional National Guard forces were managed by the National Guard Bureau, but accepting mission request for assets was up to the individual state's governors and TAGs to approve or decline. Requests for forces were often sent out as a blanket requests for any state to accept.

These methods established a kind of pick and choose approach to National Guard mission tasking. There is no formal mechanism for the Joint Force Headquarters in the state, the DCO, or the commanders of forces to coordinate with one another and have a common understanding of either what the other forces are doing or the common operational picture (COP). For the first two days of operations, military commands did not have situational awareness of what forces the National Guard had on the ground. Military forces could not operate at full efficiency when they lacked visibility of over half the military forces in the disaster area.[24] Coordination of activities and responsibilities must then be maintained at the JFO and with the JFO Coordination Group. Maintaining an accurate and timely common operational picture of military forces became difficult if not impossible. Neither the NRP nor the NIMS requires a supporting state to establish a liaison officer at the JFO. It then relies heavily on the State Coordinating Officer at the JFO to have a clear understanding of all National Guard activities and situations. Another issue is that the State Coordinating Officer is typically a civilian. He or she does not have a working knowledge of military procedures or terminology. In addition, the State Coordinating Officer is monitoring and coordinating resources for all the State agencies, to include those provided from outside the affected state, increasing the required span of responsibility. Placing a liaison officer from each of the state's National Guards involved to coordinate with the State Coordinating Officer can mitigate some of the communication risks, but this model adds another layer of coordination requirements.

Federalized National Guard

The next model examined by this study is the federalizing of the National Guard during DSCA operations. The President's authority to federalize the National Guard is

63

granted by *United States Code* Title 10, section 12304.[25] In general, the President can call up to 200,000 members of the National Guard into federal service during an emergency or disaster to "augment active forces for any operational mission."[26] While in federal status, the National Guard is under the command and control of the President through the Combatant Commander and receives federal pay and benefits. DSCA mission requests would follow the channels for federal resources. The DCO would monitor the availability of National Guard assets and capabilities and forward any missions or tasks for the National Guard to the Combatant Commander or JTF Commander. Once the National Guard is federalized, the Governor and the Adjutant General no longer have any command relationship with the federalized forces.

LA Riots Lessons Learned (Federalized)

The 1992 Los Angels riots began on 29 April, 1992 after a jury acquitted four white police officers of the 1991 beating, caught on video tape, of Rodney King, and African American motorist arrested for driving under the influence. The verdict ignited the already fragile racial environment in Los Angels. Within 45 minutes of the announcement an unruly crowd began to gather at the intersection of Florence and Normandie avenues. The initial police response witnessed a mob assaulting vehicles and pedestrians and smashing local shop storefronts and windows. Shortly thereafter, scenes of the violence in the streets of Los Angeles were being broadcast nationwide by the news media.[27] Outnumbered, the responding officers retreated leaving the crowd's destruction to worsen and the expanding television images to become a source of power for the spread of more violence. The result was one of the deadliest and most costly civil disturbances in US history with 54 people dead and 2,000 injured and property damage

64

estimates approaching a billion dollars.[28] Sometime after 8:30 p.m. that night, Mayor

Tom Bradley called Governor Pete Wilson and requested 2,000 National Guardsmen with

a specific mission to be determined later. By 4:00 a.m. the next morning, 2,000 California

Guardsmen were assembled in their armories awaiting orders.[29] Throughout the next day,

Los Angeles City and County officials requested another 4,000 National Guard troops,

but as Guard soldiers reported for duty and began deploying to the streets, violence

continued, and the Guard's response was perceived as too slow. On 1 May, the President

agreed to deploy federal troops to Los Angeles. With the deployment of federal troops

came the formation of JTF--LA and the federalization of the National Guard. When

Guardsmen heard that they had been federalized and active forces were on the way, the

feeling among some National Guard soldiers already deployed on the streets was that

their efforts were not recognized or appreciated.[30] By the time federal forces arrived, "the

riots were essentially over."[31] Among other reasons, the dusk to dawn curfew initiated

the first night and the presence 4,000 National Guardsmen and 9,000 Police Officers

from Los Angeles and other agencies across the state had a major impact on discouraging

the rioters within the first 36 hours.[32] Federal forces began arriving the night of 1 May,

and once the JTF--LA Headquarters was operational, it assumed full responsibility for all

troops in the operation, including the California National Guard. Its first priority was to

establish centralized command and control of all military forces.[33] Major General

Covault, the Commander of JTF--LA clearly intended for the command structure to be a

joint and equally shared task force. The command staff of JTF--Los Angeles integrated

all forces deployed in the operation. The Marine Corps handled the personnel and

logistics, and the J-3, Operations section was headed up by a U.S. Army colonel with a

Marine Corps deputy. The commander of Army Forces was Major General Hernandez, the 40th Infantry Division Commander of the California Army National Guard.[34] While there where many adverse impacts with the deployment of federal forces, the establishment of JTF--LA, and the federalization of the Guard, the unity of command of the "total force under Covault never worked better."[35] With the federalization, there were many changes according to the California National Guard. Before the establishment of JTF-LA and the federalization of the National Guard, virtually 100 percent of law enforcement support requests had been approved. Following federalization, only about 20 percent were approved.[36] The assumption typically made by many about the difference in this approval rate is attributed to the *Posse Comitatus Act*. The limitations placed on federal military forces prohibiting the performance of law enforcement operations are most often utilized in the argument against federalizing National Guard forces. Section 1385, Title 18 of the *United States Code*, commonly referred to as the *Posse Comitatus Act*. The intent behind the act is to prevent military forces of the United States from becoming a national police force. The execution of this law is perceived to include the arrest and detention of criminal suspects, search and seizure, restriction of civilian movement through the use of roadblocks and checkpoints, and the gathering of evidence for use in court.[37] However, this was not the case during the Los Angeles riots. When President Bush called for the deployment of federal forces to Los Angeles, he signed an Executive Order authorizing the use of the Armed Forces in Federal Law Enforcement to "suppress the violence...and to restore law and order."[38] There are several exceptions to the *Posse Comitatus Act*; the most notable is the Insurrection Act. Sections 331 and 332, Title 10 of the *United States Code*, authorize the President to use federal armed forces to enforce the laws of any state whenever there is an act of insurrection against the government.[39] The problem during the riots was that JTF--LA refused to perform law enforcement missions. The report by William H. Webster to the board of Police Commissioners on the Civil Disorder in Los Angeles, 1992 identified the issue as "The (Joint Task Force)... required each request for assistance to be subjected to a nebulous test to determine whether the requested assignment constituted a law enforcement or a military function."[40] Each new request for support was reviewed by the JTF--LA commander, the operations officer, and the staff judge advocate.[41] Additionally; JTF--LA realigned the boundaries which forces would operate in. The command used freeways to define most of the JTF boundaries. Although they provided an identifiable line on a map and were easily located on the ground, the freeways had no political or operational relevance. As a result, units who were aligned to support a single police jurisdiction were now supporting areas that often incorporated more than one police district or operational area. In some cases, the change caused units to operate in more than one city. The effects were the same from the police perspective. Police department leaders, who previously had only one military counterpart, now were typically supported by multiple units and were thus required to coordinate with more than one military headquarters.[42]

This analysis of the Los Angeles riots identifies advantages with this command relationship model. One advantage is that it ensures unity of command. The federalization of National Guard forces and the establishment of a JTF provided a single chain of command for all military forces conducting DSCA operations. At the higher level, it mitigated the issues of coordination and utilization of resources potentially caused by separate command relationships and multi--level governmental authorities.

Mission requests were sent up through the organization to a single approving authority. This model provides a means to increase the efficiency of the military response efforts during catastrophic disasters. However, it removes a large and valuable resource from the Governor's available state assets.[43]

This study also identified some disadvantages. Once called into federal service, the National Guard is subject to the same restrictions under the *Posse Comitatus Act* as active forces.[44] In most disaster circumstances, this limits the scope of missions federal or federalized forces can support for local authorities, for example establishing road blocks or check points to limit or route traffic. This was not the case in Los Angeles. At the request of the Governor, the President invoked his authority to use federalized forces found in the Insurrection Act in order to suppress insurrection against the state government.[45] Conditions to enact the provisions of the Insurrection Act are very specific and do not normally apply during conditions caused by natural disasters. This caused confusion with Los Angeles police authorities. Missions for National Guard soldiers who had recently performed law enforcement operations in a State active duty status were now under the approval authority of federal forces. JTF--LA operated under a different set of approval criteria than the National Guard was under previously. The staff of the JTF was unfamiliar with the geographic and municipality organizations and authorities. The realignment of operational boundaries was done for geographic purposes alone. It failed to take into account the organizational structures or operational areas of the civil authorities being supported.

Dual Hat Title 10/32

The Dual Hat model establishes a single chain of command for both active Title 10 and National Guard Title 32 forces. The National Defense Authorization Act of 2004, amended section 325 of Unites States Code Title 32, allowing that a National Guard officer not be released from duty in the National Guard when called into active service while serving in command of a National Guard unit. This condition must have both the consent of the Governor and the President's authorization for service in both statuses. Section 315 of *United States Code* Title 32 also allows for the detailing of active army and air force officers to the National Guard and, subject to the President's permission, the officer may accept a commission in the Army or Air National Guard of the state or territory.[46] This command relationship allows the ability to create a single JTF with both active and National Guard forces operating in an area under one command structure. The JTF--state can act as a subordinate Command and Control headquarters for USNORTHCOM if required.[47]

Operation Winter Freeze Lessons Learned (Dual Hat)

From November 2004 through January 2005, the active Title 10 and National Guard Title 32 forces participated in Operation Winter Freeze. During this operation the Border Patrol was the lead agency, and the military kept suspected terrorists out of the country. The primary mission was to detect, deter, and monitor suspicious actions using air assets.[48] The National Guard attained unity of command for all military forces operating in support of a major event. It conducted command and control from one Joint Force Headquarters, under the command of a single National Guard officer. The JTF commander had operational control over all National Guard units from multiple States operating under Title 32 authority, as well as active Title 10 forces in a joint, intergovernmental and interagency environment.[49] In addition to Operation Winter Freeze, National Guard and active forces conducted Defense Support to Civil Authorities (DSCA) operations during 2004 the at the G--8 Summit Conference, the Democratic National Convention and the Republican National Convention. The Dual Hat command relationship in each one of these operations allowed a command and control structure that was executed from a single JTF Headquarters.[50]

The advantages of the command relationships during Operation Winter Freeze were the ability to establish and maintain a unity of command. This unity of command provided a single authority and control over military activities during operations, as well as a single coordinating point of contact with US Customs and Border Patrol for coordination and support requests. Another advantage of this command relationship was that by maintaining the Title 32 status of the National Guard; military forces were able to conduct law enforcement operations. During the mission, the National Guard exposed three terrorist smuggling organizations.[51] As identified earlier National Guard forces operating under Title 32 status are not subject to the restrictions of the *Posse Comitatus Act*. This single chain of command allowed the Dual Hat Title 10, Title 32 commander to direct National Guard Title 32 forces to conduct law enforcement operations. When this command relationship is established with a National Guard Dual Hat Commander it provides a structure in which the commander has a familiar knowledge of the geography of the affected area, as well as the authorities, personnel, and structure of the local and state emergency management systems.

A possible disadvantage that was not presented in any research on the operation is in the administrative and logistic support channels. The organizational structure established by this model provides both active and National Guard forces the ability to maintain and coordinate habitual administrative and support channels, eliminating the need to merging of National Guard records into the federal system. However, it creates an additional coordination requirement and the need to maintain parallel administrative and logistic management and tracking systems.

Summary

The most import lesson from the analysis of the various command relationships is that both active and National Guard forces have conducted operations under different command relationships. What works well in one situation may not work so well in the next. The State JOC, the POMSO, and National Guard Commanders routinely provide DSCA as a resource of the State and in a State active duty status. The POMSO and JFHQ JOC staff work regularly with personnel at the Office of Emergency Services and are accustomed to the policies and procedures within their state. USNORTHCOM and federal forces are used to the organizations of the NRP and their command relationships with active higher headquarters.

---

[1]United States, Department of the Army, *How the Army Runs: Senior Leaders Reference Handbook, 2005-2006,* 25th ed. (Carlisle, PA: U.S. Army War College, 2005), 480.

[2]State of California, Governor's Office of Emergency Services, "State of California Emergency Plan," 4.

[3]Ibid., 5-6.

[4]Ibid., 18-19.

71

[5]United States, Department of the Army, "How the Army Runs: Senior Leaders Reference Handbook 2005-2006," 472.

[6]President George W. Bush, "Homeland Security Presidential Directive/HSPD-5," 3.

[7]United States Government, Department of Homeland Security, "National Response Plan," 16-17, 28.

[8]United States, Department of the Army, "How the Army Runs: Senior Leaders Reference Handbook 2005-2006," 479-480.

[9]Ibid., 480-481.

[10]United States, Department of the Army, *Defense Support to Civil Authorities, DCSINT Handbook no 1.04,* (Fort Leavenworth, KS: US Army Training and Doctrine Command, 2005), 5.

[11]United States Government, Department of Homeland Security, "National Response Plan," 34-35.

[12]LTG H. Steven Blum, "National Guard Homeland Defense Whitepaper: September 11, 2001, Hurricane Katrina, and Beyond," 4.

[13]United States Government, Department of Homeland Security, "The Federal Response to Hurricane Katrina: Lessons Learned," 34-35.

[14]LTG H. Steven Blum, "National Guard Homeland Defense Whitepaper: September 11, 2001, Hurricane Katrina, and Beyond," 4.

[15]United States Government, Department of Homeland Security, "The Federal Response to Hurricane Katrina: Lessons Learned," 26.

[16]Ibid., 42.

[17]Scott W. Johnson, *California National Guard, Joint Task Force - California, After Action Report - OPORD Gulf Coast Relief, 31 Aug - 2 Oct, 2005* (Sacramento, CA: California National Guard, 2005), 1

[18]Kochems, "Military Support to Civilian Authorities: An Assessment of the Response to Hurricane Katrina," 4.

[19]United States Government, Department of Homeland Security, "The Federal Response to Hurricane Katrina: Lessons Learned," 55.

[20]Lt Col Jeffery Richard, "Katrina Relief Deployment (CA National Guard at Armstrong International Airport New Orleans (MSY)," *Trip Report* (2005): 3

[21]Johnson, "California National Guard, Joint Task Force - California, After Action Report - OPORD Gulf Coast Relief, 31 Aug - 2 Oct, 2005,", 10

[22]United States Government, Department of Homeland Security, "The Federal Response to Hurricane Katrina: Lessons Learned," 43.

[23]United States Government Accounting Office, "Hurricane Katrina: Better Plans and Exercises Need to Guide the Military's Response to Catastrophic Natural Disasters," 9.

[24]United States Government, Department of Homeland Security, "The Federal Response to Hurricane Katrina: Lessons Learned," 55.

[25]Cornell University, The Legal Information Institute, "U.S. Code Collection,"

[26]Elizabeth B. Bazan, *Robert T. Stafford Disaster Relief and Emergency Assistance Act: Legal Requirements for Federal and State Roles in Declarations of an Emergency Or a Major Disaster* (Washington D.C.: Congressional Research Service, 2005), 9.

[27]Christopher M. Schnaubelt, "Lessons in Command and Control from the Los Angeles Riots," *Parameters, the US Army's Senior Professional Journal* Vol. XXVII, no. Summer, 1997 (2007), 92

[28]"The Flawed Emergency Response to the 1992 Los Angeles Riots," accessed 15 May 2007; available from http://www.semp.us/biots/biot_142.html.

[29]Delk, "Fires and Furies: The Los Angeles Riots of 1992," 37, 47.

[30]"The 1992 Los Angeles Riots, Military Operations in Los Angeles, 1992," in California State Military Department, The California Military Museum [database online]. Sacramento, CA 05 May, 2007, accessed 24 May 2007; available from http://www.militarymuseum.org/HistoryKingMilOps.html.

[31]Schnaubelt, "Lessons in Command and Control from the Los Angeles Riots," March 28 2007, 98.

[32]Ibid., 98.

[33]Ibid., 98.

[34]Delk, "Fires and Furies: The Los Angeles Riots of 1992," 113-114.

[35]"The Flawed Emergency Response to the 1992 Los Angeles Riots," 15 May 2007; available from http://www.semp.us/biots/biot_142.html.

[36]Schnaubelt, "Lessons in Command and Control from the Los Angeles Riots," March 28 2007, 98.

[37]"The Myth of Posse Comitatus," Sept 13 2005, accessed 24 May 2007; available from http://www.homelandsecurity.org/newjournal/articles/trebilcock.htm.

[38]Delk, "Fires and Furies: The Los Angeles Riots of 1992," 111.

[39]Cornell University, The Legal Information Institute, "U.S. Code Collection,"

[40]Delk, "Fires and Furies: The Los Angeles Riots of 1992," 305.

[41]Schnaubelt, "Lessons in Command and Control from the Los Angeles Riots," March 28 2007, 99.

[42]Ibid., 98.

[43]Bazan, "Robert T. Stafford Disaster Relief and Emergency Assistance Act: Legal Requirements for Federal and State Roles in Declarations of an Emergency Or a Major Disaster," 10.

[44]Cornell University, The Legal Information Institute, "U.S. Code Collection,"

[45]Bazan, "Robert T. Stafford Disaster Relief and Emergency Assistance Act: Legal Requirements for Federal and State Roles in Declarations of an Emergency Or a Major Disaster," 9.

[46]Cornell University, The Legal Information Institute, "U.S. Code Collection,"

[47]LTG H. Steven Blum, "National Guard Homeland Defense Whitepaper: September 11, 2001, Hurricane Katrina, and Beyond," 11.

[48]Ibid., 6.

[49]Ibid., 11.

[50]Ibid., 6.

[51]Ibid., 6.

CHAPTER 5

CONCLUSION AND RECOMMENDATIONS

While both active and National Guard forces have simultaneously conducted

Defense Support Civil Authorities operations in California in the past, neither the State,

nor the California National Guard have conducted operations with both forces since the

establishment of USNORTHCOM or the publication of the NRP or the NIMS. Following

every major or catastrophic disaster, authorities have reviewed and analyzed emergency

response at all levels. Often the lessons learned from these events have required that we

reorganize and restructure our emergency management plans and systems, always with a

single goal in mind: to improve response and recovery in order to save lives and mitigate

damages caused by disasters. As federal, state, and local authorities strive to improve our

government's ability to respond to the next catastrophic emergency, military forces must

also improve their proficiency in providing unity of effort during DSCA operations

involving both active and National Guard forces.

The establishment of USNORTHCOM after 11 September 2001 was a step in that

direction at the national level, yet an examination of the response during Hurricane

Katrina showed that the coordination and execution of a military unity of effort proved to

be difficult.[1]

The purpose of this research is to identify the most effective command

relationship between active and National Guard forces in order to provide unity of effort

in response to a catastrophic disaster within California. This chapter presents the findings

determined from the examination of command relationships in the case studies. The

conclusion takes into account these findings and the analysis of the laws, policies, systems, and concepts of emergency management in California. The final portion of this chapter and study recommends areas for further research and consideration.

During Hurricane Katrina active and National Guard forces operated under separate chains of command. Active Title 10 forces operated under JTF--Katrina, while National Guard forces operated under JTF Headquarters in both Louisiana and Mississippi in either a State active duty or Title 32 status. These separate chains of command violated the basic military principle of unity of command and made unity of effort difficult. For the first two days of operations, military commands did not have situational awareness of what forces were on the ground often leading to double tasking and duplication of effort. Military forces could not operate at full efficiency when it lacked visibility of over half the military forces in the disaster area.[2] Besides the initial search and rescue missions, military forces were needed to provide security missions in evacuated areas. The constraints of Posse Comitatus on federal Title 10 forces restricted the active force's ability to conduct these operations. Title 10 forces were unable to provide the security requirements to meet the local officials' needs. National Guard and active forces coordinated at the lower levels to establish a division of labor that would meet the needs and intent of the mission. Both National Guard and active forces were conducting the same mission because there was no central point or unity of command to coordinate activities and provide a unity of effort.

On the third day of the Los Angeles riots, the President deployed active forces to the area, at the request of the Governor and local officials. California National Guard forces already activated and deployed in response to the emergency were federalized, and JTF--LA was established to command and control both active and National Guard forces.

76

The establishment of JTF--LA provided unity of command for military activities in response to the emergency. However, the execution of DSCA operations under this command relationship presented several problems in providing effective unity of effort of all military assets to the Incident Commander. Requests for assistance were scrutinized, and it took much longer to determine the legal applications of an operation involving the use of federal forces in law enforcement missions. Active forces on the JTF--LA staff unfamiliar with the operational area and procedures of the local law enforcement authorities and agencies realigned the military's operational boundaries. The results often required law enforcement officials to coordinate with two or more military leaders supporting operations in their area. Prior to federalization, California National Guard forces rapidly approved all requests for assistance to perform their law enforcement mission. The leadership of the California National Guard was aware that under State active duty or Title 32 status, California National Guard forces were not constrained by the *Posse Comitatus Act* from conducting law enforcement missions. This allowed the mission approval process to function rapidly allowing California National Guard forces in a State active duty status to execution the mission requirement of the Incident Commander. The California National Guard's leadership and Emergency Operations Center had worked extensively in the past with both, officials at the State Office of Emergency Services, and Los Angeles Police and Sheriffs' offices. This habitual and familiar relationship allowed the California National Guard to more effectively work with state and local officials in assigning forces and a coordination of the overall goals and efforts of the Incident Commander.

The Dual--Hat chain of command established during Operation Winter Freeze provided unity of command by establishing one Joint Force Headquarters, under the command of a single commander. This single chain of command allowed the JTF Commander to coordinate and direct operations for all military forces, both active Title 10 and National Guard Title 32 forces from multiple states. Command issues revolving around command relationships between National Guard Title 32 forces were resolved prior to operations through the Emergency Management Assistance Compact (EMAC). The JTF was commanded by a National Guard officer, thus achieving unity of command. This arrangement provided a single authority over military activities during operations, as well as a single coordinating point of contact with the US Customs and Border Patrol, which allowed for unity of effort in coordination and supporting requests and activities. This single chain of command allowed the Dual Hat Title 10/32 commander to direct National Guard Title 32 forces to conduct law enforcement operations while commanding and directing the appropriate utilization of active Title 10 based on capabilities and assets. This command relationship also provided a military leadership familiar with the geography of the affected area, and a working knowledge of the authorities, personnel, and structure of the local and state emergency management systems.

Conclusion

California has a well--developed plan and a tried system to respond to disasters using the abundance of response resources available from the state and through its county mutual aid regions. The California National Guard conducts several missions in support of these emergencies under the State Emergency Management System, and continually coordinates with the Office of Emergency Services in planning, exercising, and conducting response operations. Additionally, the JFHQ--CA JOC has coordinated and directed Title 32 military assets from outside the state in responding to wild-land fires in support of the U.S. Forest Service, and the California Department of Forestry. A disaster similar in nature to what was seen during Hurricane Katrina will require response and recovery operations for a variety of missions, to include law enforcement activities, which require the capabilities and assets of both active and National Guard forces. The integration and a unified command and control of these forces are essential in providing the incident commander with unity of effort. The Dual--Hat command relationship provides the most effective model to ensuring unity of effort in emergency response operations between active and National Guard forces within California. This model provides the unity of command necessary to effectively coordinate and direct military operations, while providing the JTF commander the flexibility to use the appropriate capabilities and assets of both the active and National Guard to meet the mission requirements of the Incident Commander. The working relationships between the Office of Emergency Services and the POMSO and leadership of the California National Guard provide an understanding of the policies, procedures, and inner workings of both organizations that can efficiently and effectively coordinate the response activities of military forces and capabilities. The experience the California National Guard has with disaster response and coordinating outside Title 32 military assets provides a base for developing the Dual--Hat command relationship with active federal forces.

Recommendations

In California, when a disaster requires the use of both active and National Guard capabilities, the Department of Defense should maintain the relationships between the Office of Emergency Services and the California National Guard POMSO and leadership to provide an effective command and control system for the employment of military forces within the state. The President and Governor should enact the provisions authorized in section 325 of Unites States Code Title 32 to allow a National Guard officer to retain his or her commission in National Guard when called into active duty. With this authority established, the California National Guard should organize a JTF Headquarters with both active and National Guard staff members to coordinate a cooperative and unified response.

USNORTHCOM should lead the coordination of planning efforts and work with the California National Guard to establish agreed upon command relationships. Both USNORTHCOM and the California National Guard should describe and direct this relationship in procedures and policy. USNORTHCOM, through its coordinating authority, should facilitate the production of supporting plans with the California National Guard, National Guard Bureau, and active units allocated for disaster response incorporating the command relationship between forces and the JTF command.

Effective response to emergency operations requires extensive coordination, planning and exercising. California's Office of Emergency Services should coordinate multi-level exercises that validate the command relationship documented in military plans and the integration of the JTF into response operations.

80

Recommendations for further research

Different types of disasters will require different levels of response, capabilities, and command relationships. Further research should be done to identify the types and magnitude of disasters that require different command relationships. For example; a pandemic influenza epidemic that affects California will also affect other areas of the country. Unity of effort in a national level disaster covering multiple states, where the lead agency and unified command is at the federal level may require a different command authority.

The National Guard is the first responder for domestic military assets. Further research should examine the capabilities, organizations, and response timelines of the National Guard in each state to meet these procedures. The research should examine the creation of a standing National Guard JTF for disaster response, analyzing personnel and equipment requirements for a variety of capabilities and responses.

Summary

This thesis provides the reader with a basic understanding of the laws and policies of the emergency management systems and disaster response plans involved with DSCA operations in California. The conclusion derived from the analysis in the study provides a model for the command relationship between active and California National Guard forces that improves unity of effort in disaster response.

---

[1] J. Emery Midyette Jr., "Resource and Structure of States' National Guard," 16.

[2] US Government, Department of Homeland Security, "The Federal Response to Hurricane Katrina: Lessons Learned," 55.

GLOSSARY

California Emergency Organization. The civil government organized and augmented or reinforced during an emergency by auxiliaries, volunteers, persons pressed into service, the private sector, and community based organizations.

Defense Support to Civil Authorities (DSCA). DOD support provided by Federal military forces, DOD civilians and contract personnel, and DOD agencies and components in response to requests for assistance during domestic incidents. These incidents include terrorist threats or attacks, major disasters, and other emergencies. Although not yet in official DOD strategy or policy documents, Defense Support to Civil Authorities (DSCA) is replacing the term Military Assistance to Civil Authorities (MACA).

Emergency Management. The provision of overall operational control or coordination of emergency operations at each level of the California Emergency Organization, whether by the actual direction of field forces or by the coordination of joint efforts of governmental and private agencies.

Emergency Management Assistance Compact (EMAC). Is a congressionally ratified organization that provides form and structure to interstate mutual aid. Through EMAC, a disaster impacted state can request and receive assistance from other member states quickly and efficiently, resolving two key issues upfront: liability and reimbursement.

Emergency Support Functions (ESF). The NRP applies a functional approach that groups the capabilities of Federal departments and agencies and the American Red Cross into ESFs to provide the planning, support, resources, program implementation, and emergency services that are most likely to be needed during Incidents of National Significance. The ESFs serve as the coordination mechanism to provide assistance to State, local, and tribal governments or to Federal departments and agencies conducting missions of primary Federal responsibility.

Federal Agency (Federal definition).Any department, independent establishment, government corporation, or other agency of the Executive Branch of the Federal Government, including the United States Postal Service, but not the American Red Cross.

Federal Assistance (Federal definition). The aid to disaster victims or State or local governments by federal agencies under the provisions of the Federal Disaster Relief Act (P.L. 93-288) and other statutory authorities of federal agencies.

Federal Coordinating Officer (Federal definition). The person appointed by the President to coordinate federal assistance following an emergency or major disaster declaration.

Incident Commander (IC). The Incident Commander is the individual responsible for all incident activities, including the development of strategies and tactics and the ordering and the release of resources. The IC has overall authority and responsibility for conducting incident operations and is responsible for the management of all incident operations at the incident site.

Incident Command Post (ICP). The field location at which the primary tactical-level, on-scene incident command functions are performed. The ICP may be co-located with the incident base or other incident facilities and is normally identified by a green rotating or flashing light.

Incident Command System (ICS). A standardized on-scene emergency management construct specifically designed to provide for the adoption of an integrated organizational structure that reflects the complexity and demands of single or multiple incidents, without being hindered by jurisdictional boundaries. ICS is the combination of facilities, equipment, personnel, procedures, and communications operating within a common organizational structure, designed to aid in the management of resources during incidents. It is used for all kinds of emergencies and is applicable to small as well as large and complex incidents. ICS is used by various jurisdictions and functional agencies, both public and private, to organize field-level incident management operations.

Military Assistance to Civil Authorities (MACA). The broad mission of civil support consisting of the three mission subsets of military support to civil authorities, military support to civilian law enforcement agencies, and military assistance for civil disturbances.

Military Assistance to Civil Disturbances (MACDIS). A mission of civil support involving Department of Defense support, normally based on the direction of the President, to suppress insurrections, rebellions, and domestic violence, and provide federal supplemental assistance to the states to maintain law and order.

Military Support to Civil Authorities (MSCA). A mission of civil support consisting of support for natural or man-made disasters, chemical, biological, radiological, nuclear, or high-yield explosive consequence management, and other support as required.

Military Assistance to Civil Law Enforcement Agencies (MSCLEA). DOD activities and measures to assist federal, state and local law enforcement agencies (LEA) in counter-drug, counterterrorism, and other law enforcement operations such as security for special events to included designated National Special Security Events (NSSE).

Mutual Aid. A statewide system, developed under the authority of the California Emergency Services Act, designed to ensure that adequate resources, facilities, and other support are provided to jurisdictions whenever their own resources prove to be inadequate to cope with a given situation.

National Incident Management System (NIMS). A system mandated by HSPD-5 that provides a consistent nationwide approach for Federal, State, local, and Tribal governments; the private sector, and NGOs to work effectively and efficiently together to prepare for, respond to, and recover from incidents, regardless of cause, size, or complexity. To provide for interoperability and compatibility among Federal, State, local, and Tribal capabilities, the NIMS includes a core set of concepts, principles, and terminology. HSPD-5 identifies these as the ICS; multi-agency coordination systems; training; identification and management of resources (including systems for classifying types of resources); qualification and certification; and the collection, tracking, and reporting of incident information and incident resources.

National Response Plan (NRP). A plan mandated by HSPD-5 that integrates Federal domestic prevention, preparedness, response, and recovery plans into one all-discipline, all-hazards plan.

National Special Security Events (NSSE). Events of national significance that require greater visibility.

Office of Emergency Services. Part of the Governor's office, the primary State agency responsible for the coordination and administration of statewide operations to support emergency mitigation, preparedness, response, and recovery activities within California.

Standardized Emergency Management System (SEMS). The Standardized Emergency Management System is the group of principles for coordinating State and local emergency response in California. SEMS provides for a multiple level emergency response organization and is intended to facilitate the flow of emergency information and resources within and between the organization levels.

Unified Action. A broad generic term that describes the wide scope of actions (including the synchronization of activities with governmental and nongovernmental agencies) taking place within Unified Commands, subordinate Unified Commands, or joint task forces under the overall direction of the commanders of those commands.

BIBLIOGRAPHY

Bazan, Elizabeth B. *Robert T. Stafford Disaster Relief and Emergency Assistance Act: Legal Requirements for Federal and State Roles in Declarations of an Emergency or a Major Disaster.* Washington, DC: Congressional Research Service, 2005.

Bell, Raymond E. Jr. "U.S. Northern Command and the National Guard." *Joint Forces Quarterly*, no. 36 (2005): 36-40.

Blum, H. Steven, LTG. *National Guard Homeland Defense Whitepaper: September 11, 2001, Hurricane Katrina, and Beyond.* Washington, DC: National Guard Bureau, 2005.

Brasch, Walter M. *'Unacceptable': The Federal Response to Hurricane Katrina.* Charleston, SC: BookSurge, 2006.

"Homeland Security Presidential Directive/HSPD-5." In Office of the Press Secretary database online. Washington, DC: 28 February 2003 cited 2007. Available from http://www.whitehouse.gov/news/releases/2003/02/20030228-9.html.

Childs, John Brown. *Hurricane Katrina Response and Responsibilities.* Edited by John Brown Childs. Santa Cruz, CA: New Pacific Press, 2005.

Chong, Jia-Rui, and Hector Becerra, "Katrina's Aftermath; California Earthquake could be the Next Katrina," *Los Angeles Times,* (8 September 2005): A 1.

Cooper, Christopher, and Robert Block. *Disaster: Hurricane Katrina and the Failure of Homeland Security.* 1st ed. New York, New York: Time Books, Henry Holt and Company, 2006.

Cornell University, Law School "U.S. Code Collection." database online 30 May 2006 cited 2007. Available from http://www4.law.cornell.edu/uscode/.

Davis, Lynn E., David E. Mosher, Richard R. Brennan, Michael D. Greenberg, Scott McMahon, and Charles W. Yost. *Army Forces for Homeland Security.* Santa Monica, CA: RAND Corporation, 2004.

"The 1992 Los Angeles Riots, Military Operations in Los Angeles, 1992." In California State Military Department, The California Military Museum. database online. Sacramento, CA 5 May 2007 cited 2007. Available from http://www.militarymuseum.org/HistoryKingMilOps.html.

Delk, James D. *Fires and Furies: The Los Angeles Riots of 1992.* Palm Springs, CA: ETC Publications, 1995.

Dunlap, Charles J., Jr., "Putting Troops on the Beat," *The Washington Post,* 30 September 2006.

Flynn, Stephen. *The Edge of Disaster Rebuilding a Resilient Nation.* New York: Random House, 2007.

Gecowitz, Greg, Lt Col, USAF. "Coordination, Command, Control, and Communications." *Joint Center for Operational Analysis, Quarterly Bulletin* VIII, no. 2 (June 2006): 16-24.

Haddow, George D., and Jane A. Bullock. *Introduction to Emergency Management.* 2nd ed. Burlington, MA: Elservier Butterworth-Heinemann, 2006.

Johnson, Scott W. *California National Guard, Joint Task Force – California: After Action Report - OPORD Gulf Coast Relief, 31 Aug - 2 Oct, 2005.* Sacramento, CA: California National Guard, 2005.

Khomchenko, Sergey. "Civil-Military Relations in Domestic Support Operations: The California National Guard in Los Angeles 1992 Riots and Northridge Earthquake of 1994." Ph.D. diss., Naval Postgraduate School, 1997.

Kochems, Alane. "Military Support to Civilian Authorities: An Assessment of the Response to Hurricane Katrina." *Backgrounder*, No. 1899 (28 November 2005): 1-6.

Kunreuther, Howard, and Mark Pauly. "Rules rather than Discretion: Lessons from Hurricane Katrina." *Journal of Risk and Uncertainty* 33, no. 1-2 (September 2006): 101.

Larson, Eric V., and John E. Peters. *Preparing the U.S. Army for Homeland Security: Concepts, Issues, and Options.* Santa Monica, CA: RAND Corporation, 2001.

Lowenberg, Timothy J. "The Role of the National Guard in National Defense and Homeland Security." *National Guard* 59, no. 9 (September 2005): 97.

McDonnell, Janet A. *Response to the Loma Prieta Earthquake.* Fort Belvoir, VA: U.S. Government Printing Office, 1993.

Midyette, J. Emery, Jr. "Resource and Structure of States' National Guard." *Joint Center for Operational Analysis, Quarterly Bulletin* VIII, no. 2 (June, 2006): 25-42.

Miskel, James F. *Disaster Response and Homeland Security what Works, What Doesn't.* Westport, CT: Praeger Security International, 2006.

Reiterman, Tim, "Q&A/HENRY RENTERIA: Making Sure State is Ready in Case Disaster Strikes" *Los Angeles Times,* (12 September 2005): B 2.

Richard, Jeffery, Lt Col. "Katrina Relief Deployment (CA National Guard at Armstrong International Airport New Orleans (MSY)." *Trip Report* (2005).

Schnaubelt, Christopher M. "Lessons in Command and Control from the Los Angeles Riots." *Parameters, the US Army's Senior Professional Journal* Vol. 27, (summer 1997): 88-109.

Scott, Robert Travis, "Guard Leaders Adjust Storm Response ; Officials of 12 States Coordinate Efforts," *Times - Picayune,* (18 May 2006): 15.

Starr, Kevin, "California's Calamity in Waiting," *Los Angeles Times,* (20 November 2005): M 1.

State of California, Governor's Office of Emergency Services. *State of California Emergency Plan.* Sacramento, CA: Office of Emergency Services, Planning Section, 2005.

Suburban Emergency Management Project, "The Flawed Emergency Response to the 1992 Los Angeles Riots." 21 November 2004. database online cited 2007. Available from http://www.semp.us/biots/biot_142.html.

"Official California Legislative Information, California Codes Military and Veterans Code." 1 January 2007 cited 2007. Available from http://leginfo.public.ca.gov.

The Gale Group, Inc, "West's Encyclopedia of American Law, "National Guard".1998. cited 2007. Available from http://www.answers.com/topic/united-states-national-guard.

"The Myth of Posse Comitatus." 13 September 2005 cited 2007. Available from http://www.homelandsecurity.org/newjournal/articles/trebilcock.htm.

United States Government Accounting Office. *Hurricane Katrina: Better Plans and Exercises Need to Guide the Military's Response to Catastrophic Natural Disasters.* Washington, DC: U.S. Government Printing Office, 2006.

United States Government, Department of Homeland Security. *National Incident Management System (Draft).* Washington, DC: U.S. Government Printing Office, 2007.

United States Government, Department of Homeland Security. *The Federal Response to Hurricane Katrina: Lessons Learned.* Washington, DC: U.S. Government Printing Office, 2006.

United States Government, Department of Homeland Security. *National Response Plan.* Washington, DC: U.S. Government Printing Office, 2004.

"Unified Command Plan." 3 May 2007 cited 2007. Available from
    http://www.defenselink.mil/specials/unifiedcommand/.

United States, Department of Defense. Joint Publication 3-0, *Doctrine for Joint
    Operations.* Washington, DC: United States Joint Forces Command, Director for
    Operations, 2006.

United States, Department of Defense. Joint Publication 1-02, *Department of Defense
    Dictionary of Military and Associated Terms*
    Washington, DC: Directorate for Operational Plans and Joint Force Development,
    2001.

United States, Department of Defense. Joint Publication 1, *Joint Warfare of the Armed
    Forces of the United States*
    Washington, DC: Directorate for Operational Plans and Joint Force Development,
    2000.

United States, Department of the Army. *Defense Support to Civil Authorities,* DCSINT
    Handbook no 1.04. Fort Leavenworth, KS: US Army Training and Doctrine
    Command, 2005.

United States, Department of the Army. *How the Army Runs: Senior Leaders Reference
    Handbook 2005-2006.* 25 ed. Carlisle, PA: U.S. Army War College, 2005.

United States, Department of the Army. Field Manual 6-0, *Mission Command: Command
    and Control of Army Forces.* Washington, D.C.: Headquarters, Department of the
    Army, 2003.

Wermuth, Michael A. *Enhancing Emergency Preparedness in California: Testimony
    Presented to the Little Hoover Commission January 26, 2006.* Santa Monica, CA:
    RAND Corporation, 2006.